T0234456

Giant Resonances

Contemporary Concepts in Physics

A series edited by

Herman Feshbach
Massachusetts Institute
of Technology

Founding Editor

Henry Primakoff
(1914–1983)

Associate Editors

Mildred S. Dresselhaus
Massachusetts Institute
of Technology

Mal Ruderman
Columbia University

S.B. Treiman
Princeton University

This book is part of a series. The publisher will accept continuation orders which may be cancelled at any time and which provide for automatic billing and shipping of each title in the series upon publication. Please write for details.

Giant Resonances

Nuclear Structure at Finite Temperature

P. F. Bortignon

and

A. Bracco

Università di Milano and *Istituto Nazionale di Fisica Nucleare, Italy*

R. A. Broglia

Università di Milano and *Istituto Nazionale di Fisica Nucleare, Italy*

Niels Bohr Institute, University of Copenhagen, Denmark

CRC Press
Taylor & Francis Group
Boca Raton London New York

CRC Press is an imprint of the
Taylor & Francis Group, an **informa** business

CRC Press
Taylor & Francis Group
6000 Broken Sound Parkway NW, Suite 300
Boca Raton, FL 33487-2742

First issued in paperback 2020

© 1998 by Taylor & Francis Group, LLC
CRC Press is an imprint of Taylor & Francis Group, an Informa business

No claim to original U.S. Government works

ISBN-13: 978-0-367-45576-7 (pbk)
ISBN-13: 978-90-5702-570-9 (hbk)

Visit the Taylor & Francis Web site at
http://www.taylorandfrancis.com

and the CRC Press Web site at
http://www.crcpress.com

British Library Cataloguing in Publication Data

Bortignon, P. F.
    Giant resonances : nuclear structure at finite temperature.
    – (Contemporary concepts in physics ; v. 10)
    1.Nuclear magnetic resonance, Giant
    I.Title   II.Bracco, Angela   III.Broglia, R. A.
    538.3′62

For

Donatella, Gianandrea and *Bettina*

— AB and RAB

Inger-Marie and *Ulla Maria*

— PFB

Contents

Part 2   FINITE TEMPERATURE

Preface to the Series

The series of volumes, *Contemporary Concepts in Physics*, is addressed to the professional physicist and to the serious graduate student of physics. The subjects to be covered will include those at the forefront of current research. It is anticipated that the various volumes in the series will be rigorous and complete in their treatment, supplying the intellectual tools necessary for the appreciation of the present status of the areas under consideration and providing the framework upon which future developments may be based.

Preface

The subject of many-body systems constitutes a central chapter in the study of quantum mechanics, with applications ranging from elementary particle and condensed matter physics to the behaviour of compact stellar objects. Quantal size effects is one of the most fascinating facets of many-body physics; this is testified to by the developments taking place in the study of metallic clusters, fullerenes, nanophase materials, and atomic nuclei.

A natural way to see what an atomic nucleus looks like is to shine a beam of photons on it and determine which frequencies it absorbs. An equally natural method is to heat it up and measure the frequency of the photons it emits. In such experiments it is observed that the atomic nucleus displays resonant behaviour, absorbing or emitting essentially monochromatic photons. The cross-section associated with the resonance indicates that essentially all nucleons participate in the process — thus, the character of "giant" is ascribed to the resonant behaviour.

To study emission processes one must first heat up the nucleus, then observe the way it decays. In other words, one has to measure the γ-decay of compound nuclei (nuclei where the excitation energy is distributed uniformly over all the degrees of freedom of the system). In this way one probes the structure of the nucleus at finite temperature.

This book is divided into two main parts: the study of giant resonances based on the atomic nucleus ground state (zero temperature), and the study of the γ-decay of giant resonances from compound (finite temperature) nuclei. In keeping with the fact that mathematical models and physical concepts used to describe the atomic nucleus are closely connected to results of observations, we begin each of these two parts by reviewing experimental information. Within this context we also discuss some of the experimental devices used in the study of giant resonances. This allows us to discuss, in simple terms, a variety of seemingly contradictory findings which, properly interpreted, testify to the stability of giant resonances as a function of temperature. We then proceed to introduce the concept of mean field, the basic organizing element of all the phenomena covered in this volume. After this, we consider the main properties of the response of atomic nuclei subject to weak external fields. The consequences of the coupling of giant resonances to the nuclear surface, the main quantal size effect within the present context, and the subject of the damping phenomenon of nuclear motion are then addressed. The parallel development of these subjects at both zero and finite temperature emphasizes the simplicity and universality of the concepts that form the basis of the nuclear field description of the atomic nucleus.

Having built our case step by step, we dedicate the last two chapters to the study of the giant dipole resonance and of rotations of hot nuclei, with special emphasis on the associated relaxation mechanisms and on their sensitivity to temperature.

The material presented here is an outgrowth of lectures given to fourth-year students at the University of Milan. We have, therefore, placed special emphasis on the role played by the general physical concepts (the patrimony of a fourth-year student) that form the foundation of the phenomenon of giant resonances. We have supplemented the basic subject matter of the lectures in an organic fashion, incorporating material taken from work going on at the forefront of research on the structure of hot nuclei. Thus, this monograph should be useful to young researchers and experienced practitioners alike.

We wish to thank Gianluca Colò, Franco Camera, Jens Jørgen Gaardhøje and Adam Maj for many discussions through the years, resulting in a most fruitful collaboration. Within this context, the special role played by Erich Ormand is acknowledged here. We thank him also for comments on the manuscript of this book. Discussions with George Bertsch, Paola Bosetti, Paola Donati, Thomas Døssing, Nicola Giovanardi, Bent Herskind, Bent Lauritzen, Silvia Leoni, Marco Mattiuzzi, Ben Mottelson, Marcello Pignanelli, Nguyen Van Giai, Enrico Vigezzi and Vladimir Zelevinsky are gratefully acknowledged. Some of the artwork for this book was prepared by Elisabeth Grothe, whom we thank for her advice and help. We also express our gratitude for the inspiring atmosphere of the Department of Physics of the University of Milan (in particular, to the members of the Nuclear Physics Group), and to the Niels Bohr Institute of the University of Copenhagen, institutions where this work was carried out. The support of the Istituto Nazionale di Fisica Nucleare, Sezione di Milano is gratefully acknowledged.

1

Introduction

One of the basic aims of modern research in nuclear structure is to subject the nucleus to extreme conditions of internal excitation energy and of rotational frequency in order to learn how the nuclear motion, as known from the study of the nuclear spectra close to the ground state, is modified by the temperature and the angular momentum content of the system. Extreme conditions of internal excitation energy mean, within the present context, temperatures above which the nucleus undergoes a liquid–gas transition ($T \sim$ 5–7 MeV; cf. Bertsch and Siemens (1983)). Extreme rotational frequencies imply rotational frequencies of the order of those inducing fission ($\hbar\omega_{rot} \approx$ 1.0–1.5 MeV, $I \approx$ 80–100 $\hbar$; cf. Cohen et al. (1974)).

The question concerning the dependence of nuclear motion with temperature and with angular momentum is equivalent, to a large extent, to the question of how single-particle levels and collective motion are affected by the internal excitation energy of the system and by the rotation of the nucleus as a whole. The signals arising from the γ-decay of giant resonances in hot, strongly rotating nuclei are, to date, among the clearest in this quest.

1.1 Overview

Under normal conditions nuclei are in their ground state, that is at zero temperature. This is because nuclei on earth, leaving aside those which arrive in the form of cosmic rays, are isolated. In fact, for two nuclei to interact, they need to have large kinetic energies, of the order of tens of MeV, so as to be able in a collision to overcome the Coulomb repulsion and reach within the range of the nuclear attraction. Energies of such magnitude contrast with the energies available at room temperature ($\approx 25 \times 10^{-3}$ eV). This is the reason why to make two nuclei interact, one needs large machines. In these machines an atom containing a heavy atomic nucleus is stripped of most of its electrons becoming a heavy ion, which is subsequently accelerated and collimated. The resulting beam, aimed at a target of heavy atomic nuclei, will eventually lead

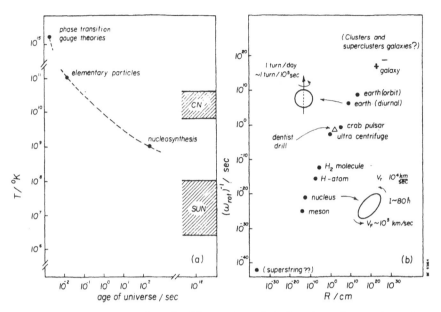

Figure 1.1. Temperatures and rotational frequencies typical of reactions where two heavy ions fuse and eventually, after a time of the order of 10^{-21} sec equilibrize all its degrees of freedom, forming a single, compound nucleus. (a) The scale of temperature has been set in connection with the temperature ascribed to the Universe at different times after the Big Bang, going through the first three minutes to our times. (b) The inverse of the rotational period (in seconds) of a variety of rotating objects observed (or thought to exist, like superstrings, the mathematical connotation of an elementary particle) in the Universe from the largest to the smallest, are quoted as a function of their radius. Both the inverse frequencies and the radii span sixty orders of magnitude, the relation between the two quantities being almost linear.

to a heavy ion reaction. In the event in which, the two nuclei fuse, the energy and angular momentum of relative motion becomes mostly excitation energy and angular momentum of the resulting composite system (cf. e.g., Broglia and Winther (1991)). Typical values of these quantities are set in Fig. 1.1, in relation to temperatures and rotational frequencies observed in other physical systems.

After the fusion process has taken place, the mean field of such a system takes a very short time to get established, namely the time needed for nucleons to exchange pions ($\approx 5 \cdot 10^{-24}$ sec). Collisions of nucleons moving in this mean field distribute the excitation energy of the fused system among all the single-particle degrees of freedom in an uniform way. Because the density of levels at even moderate excitation energies is very high ($\approx 10^{22}$/MeV for a nucleus of mass number ≈ 100 at 50 MeV of excitation energy), one can view this process as a proper thermalization of the excitation energy.

The time it takes for the system to thermalize the single-particle degrees of freedom is of the order of $2 \cdot 10^{-23}$ sec, which is the time a nucleon needs to transverse the average distance which separate nucleons. This time is the

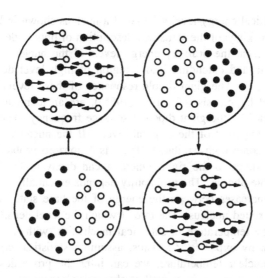

Figure 1.2. Schematic representation of the giant dipole resonance in atomic nuclei. The wavelength of the photon exciting this vibration is large with respect to the diameter of the system. Under the influence of this field the protons (filled circles) are accelerated in one direction. The neutrons (open circles) are unaffected by the field, but they move in the direction opposite to that of the protons so that the center of mass of the nucleus remains stationary and momentum is conserved. The restoring force, which ultimately reverses the motion of protons and neutrons is the strong nuclear force. The amplitude of the motion is greatly exaggerated in this figure. The actual displacement is of the order of one percent of the nuclear radius when a single quantum of vibration is present (after Bertsch and Broglia (1986)).

shortest in a hierarchy of times associated with the thermalization of the variety of nuclear degrees of freedom, a process which eventually leads to the compound nucleus as originally formulated by Niels Bohr (1936). In particular, it takes a longer time to equilibrate the collective degrees of freedom, modes in which all of the nucleons display a single, regular periodic pattern of motion, than the single-particle degrees of freedom. Giant resonances and collective rotations are expected to become in thermal equilibrium with the single-particle degrees of freedom at progressively longer times.

Giant resonances are associated with the elastic, short time response of the system. A typical example of this motion, and the only one experimentally studied to date at finite temperatures is that of the oscillations of neutrons against protons, a motion which is known as the giant dipole resonance (cf. Fig. 1.2).

The key assumption in the statistical description of nuclear decay is that all states at a given excitation energy are equally populated and that from the inverse reaction rate, one can infer the rate at which the nucleus decays by emitting a given particle. This is the essence of Niels Bohr's compound-nucleus model. It has long been used to describe nuclei with excitation energies of the order of tens of MeV, and heavy-ion reactions have extended the domain of the model to nuclei produced with excitation of hundreds of MeV.

The hypothetical energy-level scheme of a nucleus shown in Fig. 1.3 helps illustrate the workings of the statistical description of photon decay. The lines to the left represent the ordinary energy levels of the nucleus, starting from the ground state and extending into a region of high level density, at high excitation energies. The giant dipole resonance of these levels, according to the Brink–Axel hypothesis (cf. Brink (1955)), are an identical set of levels displaced upward by the giant dipole resonance frequency. These are shown in Fig. 1.3, to the right of the original levels. If the nucleus is in statistical equilibrium at some high excitation, there is a nonzero probability that it is in one of the states to the right, where it can decay to the base state by emitting a dipole photon. The probability depends essentially on how fast the level density increases between the energy of the base state and the energy of the initial excited nucleus. The rate at which the level density increases is parametrized by temperature in statistical mechanics, with the ratio of level densities given by $\exp^{-\Delta E/kT}$. Thus, assuming statistical equilibrium and knowing the nuclear temperature, we can infer the properties of the giant dipole resonance in emission as well as absorption processes.

Because of the principle of detailed balance, the time it takes for the fused system to equilibrate the variety of degrees of freedom is associated with the lifetime these modes display at a given excitation energy. Typically, a giant resonance goes through few (3–4) cycles before relaxing, while a deformed nucleus can keep its phase coherence typically for 10 rotational periods. This in keeping with the fact that typical values of the centroid and the damping width associated with the giant dipole resonance are, in hot nuclei, 15 MeV and 5–10 MeV respectively, while typical rotational transition energies and damping widths in warm nuclei are 1 MeV and 100 keV respectively. In other words, it takes $\approx 10^{-22}$ sec for a giant resonance to become in thermal equilibrium with the other degrees of freedom in the compound nucleus and $\approx 10^{-20}$ sec for the rotational motion. The study of these relaxation times, together with that associated with the single-particle degrees of freedom, occupies an important place in the present monography.

Relaxation of a many-body system is controlled by two processes: a) collisional damping arising from collisions among the particles, b) inhomogeneous damping, associated with the anisotropy displayed by the system to motion along the different directions in space. These mechanisms have, in the nuclear case, shown unexpected new facets. This is due to the central role played by the nuclear surface in the damping process (cf. Chapters 4, 9, 10 and 11). In fact, in a binary encounter between nucleons, a nucleon can change its state of motion, by promoting a particle moving in the Fermi sea into a state above the Fermi surface. Because particle–hole correlations are, in the nuclear case, considerably stronger than particle–particle correlations, as testified by the larger collectivity displayed by surface modes as compared to pairing modes (cf., e.g., Broglia et al. (1973)) and because the Pauli principle is more effective in the bulk than in the surface of the system, binary collisions can be accurately accounted for, by processes where a nucleon bounces inelastically off

Figure 1.3. Nuclear energy levels in schematic representation. The levels depicted to the left (L) are the ordinary nuclear states. The levels shown to the right (R) are giant dipole excitations. These states are built from the ordinary ones by the addition of a quantum of vibrational excitation, which raises the energy of the state by $\hbar\omega_{dipole}$. This picture is idealized in that the states are depicted as distinct, sharp levels. In reality they are broad and strongly mixed. However, this broadening and mixing is not relevant for the statistical arguments.

the surface and sets it into vibration. Concerning inhomogeneous damping, a hot nucleus can be viewed as a time-dependent ensemble of systems displaying different shapes. Collective modes probe both the associated spatial inhomogeneity of this ensemble as well as the time spent by the system in each configuration, a fact which is reflected in the distribution of frequencies of the γ-decay process.

Due to the fact that: (a) phase interference typical of quantal phenomena screens the coupling of collective modes with other degrees of freedom of the nucleus (cf. Sects. 4.2, 9.2, 10.6 and 11.5), (b) the phenomenon of motional narrowing associated with rapid thermal fluctuations of the nuclear surface reduces the importance of inhomogeneous damping (cf. Sects. 10.7, 10.8 and 11.3), it is expected that collective motion will exist in nuclei even at the highest temperatures these systems can sustain (5–7 MeV). However, because the nucleus can cool by evaporating particles as a liquid drop does, in some cases faster than the time needed for the vibrational modes to be thermally excited, it is not said that giant resonances or rotational motion can be observed at all temperatures (cf. Sect. 6.2).

The study of collective motion in hot nuclei in general, and of giant resonances in particular, provides a unique probe to study the evolution of the nuclear structure as a function of temperature. This is one of the reasons for the subtitle of the present monograph. The other is associated with the fact that the techniques developed to study giant resonances in hot nuclei, can also be used at profit to learn about other types of nuclear motion in excited nuclei, in particular rotational motion (Ch. 11).

We start, in Chapter 2, with an overview on the experimental information associated with well known giant resonances based on the ground state. Following this survey we present in Chapters 3 and 4 the theoretical tools needed to provide an overall description of the experimental findings. We then proceed to discuss the results of observations of the giant dipole resonance, only vibration detected to date in the decay of a highly excited system (Chapters 5 and 6). The tools needed to provide a theoretical understanding of these findings are developed in Chapters 7, 8 and 9. We conclude by applying these tools to the discussion of the giant dipole vibration in hot nuclei (Chapter 10) and of rotational motion in warm nuclei (Chapter 11).

In the remainder of the present chapter, we shall provide a compendium of the basic questions and well established results concerning the physics of nuclei at finite temperature and angular momentum. It may be used at profit, as a guide, in reading the following chapters.

1.2   Hot Nuclei

The connection between statistical mechanics and thermodynamics is customarily established in the limit of large A, where A is the number of particles in the system under consideration. In connection with the present discussion, one needs to ask how small can A be without the concepts of temperature and entropy losing their applicability?

Let us agree to limit the use of temperature to systems possessing an energy reservoir that can be shared statistically, and also possessing a density of states whose logarithmic derivative varies smoothly as a function of the internal excitation energy of the system over a substantial energy range. These conditions appear to be satisfied in the case of heavy nuclei excited to several tens of MeV, even though they contain a small number of particles. In fact, for a nucleus of mass number $A = 200$ at an excitation energy of 50 MeV, the Fermi gas formula gives about 10^{28} states per MeV, and the density should increase by a factor of roughly e for an increase of 1 MeV in the excitation energy. How does one make a heat reservoir in the nucleus? While it is not a thermal bath in the classical sense, when the system emits a neutron or a γ-ray in the cooling process, it exchanges energy statistically with the freed particle. This is in keeping with the fact that the energy distribution of the emitted nucleon or γ-ray is determined by the density of levels of the daughter states. Within this scenario let us assume that both the neutron and γ-ray energy spectra then fall off with a factor e for each MeV increase in energy of the emitted particle. It is then natural to interpret this behaviour as arising from a Boltzmann factor with a temperature of 1 MeV (Bertsch (1988)).

In connection with the study of giant vibrations in hot nuclei, we show in Fig. 1.4(a) a typical measured γ-ray spectrum for the decay of a compound nucleus resulting from the heavy-ion reaction $^{12}C + ^{154}Sm$ at 60 MeV of carbon-bombarding energy. The γ-emission rate falls over six orders of magnitude, as the γ-energies go from 5 to 25 MeV, and provides information on the temperature of the system. The spectrum shows an undulation, the primary evidence of the giant dipole resonance. This is clearer from Fig. 1.4(b), which is a plot of the same data rescaled by an exponential factor to roughly divide out the effects of the density of levels and the probability of formation of the dipole state.

1.3 Single-Particle Motion

The special stability ascribed to nuclei with given numbers of protons and neutrons ("magic numbers"), testifies to the fact that nucleons in nuclei move essentially independent of each other, feeling the pushings and pullings of other nucleons only when trying to leave the system. The associated mean field defines a surface which can vibrate leading to a spectrum of collective, low-lying excitations. Around closed shell nuclei, these vibrations are found within 1–3 MeV of excitation energy. For nuclei progressively removed from closed shells, and due to the polarization effects of particles outside closed shells, the frequencies of some of these vibrations, in particular of quadrupole vibrations, can become so low, that the nuclear surface becomes unstable and the nucleus acquires a static deformation. The phenomenon of "phase

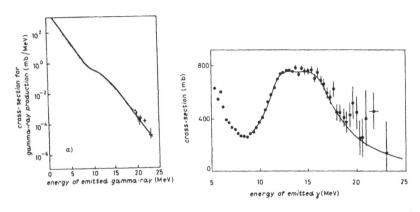

Figure 1.4. a) Measured spectrum from decay of ^{166}Er nucleus, formed in the reaction of ^{12}C and ^{154}Sm nuclei. The compound nucleus is created in a state of 50 MeV intrinsic excitation energy. b) The data shown in a) are here displayed with an exponential factor divided out. The giant dipole resonance now appears as a flat-topped peak. The solid curve in both a) and b) shows a fit to the data from a computer simulation of the statistical decay process. From Gosset et al. (1985).

transition" from a spherical to a deformed configuration is testified by the presence of vibrational and rotational bands in the nuclear spectrum, as shown in Fig. 1.5.

The coupling of the nucleons to dynamical vibrations of the surface, a process which goes beyond mean field, can strongly renormalize the single-particle motion by changing the energy and the occupancy of levels around the Fermi energy. Also, by providing a damping width Γ_{sp} and thus a finite lifetime to single-particle levels (Sect. 4.1.1). Because of the limited amount of available data, it is not possible to empirically discriminate between a linear or a quadratic dependence of Γ_{sp} on the single-particle energy as measured from the Fermi energy. However, general arguments connected with the central role played by the nuclear surface in the damping process, suggest a linear dependence.

In hot nuclei, a single-particle level at the Fermi level will also acquire a finite damping width Γ_{sp}. This is because a nucleon moving in this level, has the possibility of making real transitions to more complicated configurations which are thermally excited. Due to the fact that the temperature T can be viewed as the energy available to make such transitions, Γ_{sp} is also expected to vary linearly with temperature (Sect. 9.1.1). This behaviour is very different from the one observed in the case of infinite systems, where the damping width of single-particle motion depends quadratically on temperature (Morel and Nozieres (1962)). This difference reflects the special role the nuclear surface plays in the relaxation of nuclear motion.

The estimated values of Γ_{sp} imply that, up to temperatures where the shell structure melts, the mean free path of a nucleon at the Fermi energy is expected

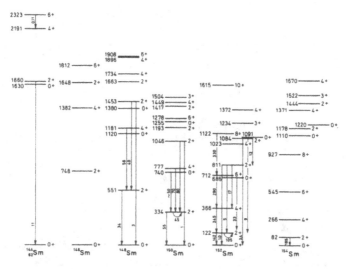

Figure 1.5. Spectra of even-parity states in even–even Sm-isotopes illustrating the shape phase transition between a spherical (A<150) and a quadrupole deformed phase (A>150). From Bohr and Mottelson (1975).

to be of the order of the nuclear dimensions (Sect. 4.1.1), and mean field theory to be useful in the description of the nuclear structure.

1.4 Giant Resonances

The giant dipole resonance corresponds to a back and forth sloshing of neutrons against protons (cf. Fig. 1.2). Its energy centroid is inversely proportional to the nuclear radius (cf. Fig. 2.5(b)), that is, it is proportional to the nuclear momentum, typical of elastic vibrations. In a spherical nucleus, the vibrations along any three perpendicular axes are equivalent, the strength of the vibration being equally distributed in the three directions. In deformed nuclei, the giant dipole strength will split. In particular, in prolate systems there will be two peaks. One associated with vibrations of protons against neutrons along the symmetry axis and one, at higher energy and carrying twice the strength of the first peak, associated with vibrations perpendicular to the symmetry axis. This is a consequence of the phenomenon of inhomogeneous damping, reflecting the spatial anisotropy of the system. An example of this phenomenon is shown in Fig. 1.6.

Angular momentum also affects the breaking of the resonance. In fact, under the strain of the associated Coriolis force the nucleus will change deformation. Because angular momenta in a compound state can be as high as 70–80 units of $\hbar$ (cf. Chs. 5 and 6; cf. also Sect. 10.8.2), the deformation induced by rotations can be sizable ($\beta \approx 0.2$–0.3) and the associated splittings conspicuous (2–3 MeV). Each of the spectroscopic lines associated with vibrations along a

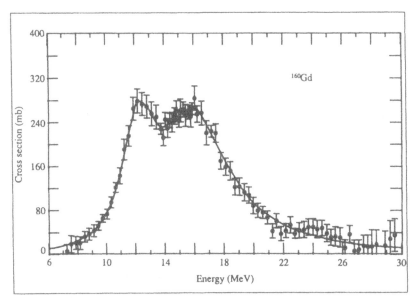

Figure 1.6. Photoabsorbtion cross section in ^{160}Gd (from Berman and Fultz (1975)).

different deformation axis will further split due to the rotation of the system (cf. Fig. 1.7). This phenomenon is well known from the spectroscopic studies of celestial rotating bodies.

Each of the resulting spectroscopic lines can undergo further splitting or eventually acquire a width, still within mean field, due to the following effects:

1. decay into single-particle motion,

2. particle-decay,

3. γ-decay.

Because giant resonances lie at energies of the order of the energy difference between major shells, they are embedded in a background of particle–hole excitations, into which they can, in principle, decay (Sect. 3.2.1). This in keeping with the fact that giant resonances can be viewed as a correlated particle above the Fermi surface and a hole in the Fermi sea. Decay of a collective vibration into uncorrelated particle–hole excitations is a phenomenon known as Landau damping in solid state physics. Furthermore, giant resonances lie above threshold for particle emission. Consequently, they can become damped by particle decay (cf. Sects. 2.5 and 4.3). They can also be damped through coupling to the electromagnetic field leading to the emission of a photon (cf. Sects. 2.6 and 4.4).

None of the three relaxation mechanisms of giant resonances listed above, are expected to be of importance for the giant dipole resonance of medium-heavy nuclei. This is because: a) the centroid of the resonance will, in general,

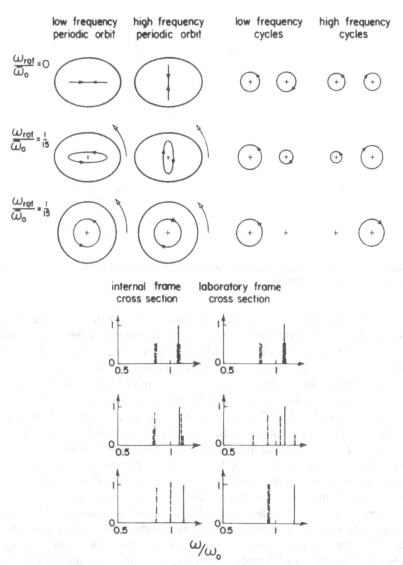

low frequency periodic orbit   high frequency periodic orbit   low frequency cycles   high frequency cycles

$\frac{\omega_{rot}}{\omega_0} = 0$

$\frac{\omega_{rot}}{\omega_0} = \frac{1}{15}$

$\frac{\omega_{rot}}{\omega_0} = \frac{1}{15}$

internal frame cross section   laboratory frame cross section

ω/ω_0

Figure 1.7. Periodic orbits in a rotating deformed harmonic oscillator potential. The left-hand part shows one equipotential contour together with the two periodic orbits in the plane perpendicular to the rotation vector for a prolate potential without (top row) and with (middle row) rotation, and for an oblate potential with the rotation vector pointing in the direction of the symmetry axis (bottom row). The size of the quadrupole deformation is 0.25. The amplitude of the orbits shown is selected to yield an energy in the rotating frame proportional to the frequency, corresponding to a specific number of phonons. In the right-hand side of the figure is shown the decomposition of the periodic orbits on cyclic motion. The absorption cross section of the electromagnetic radiation, exciting the periodic motion, is shown (in relative units) at the bottom of the figure. Dashed bars denote absorption accompanied by a decrease ($\Delta I = -1$) of the angular momentum, being proportional to the square of the radius of the clockwise cycle, and equivalently, dot-dashed bars, are associated with increase of the angular momentum ($\Delta I = +1$), given by the size of the counterclockwise cycle. Thin solid bars denote absorption associated with the linear oscillation along the angular frequency vector (after Gallardo et al. (1985)).

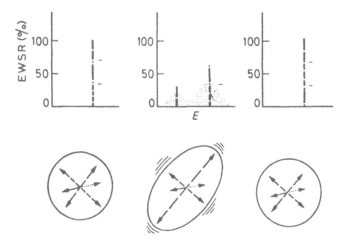

Figure 1.8. Schematic representation of the coupling of the giant dipole resonance to quantal fluctuations of the nuclear surface.

lie at excitation energies where the background of uncorrelated particle–hole excitations have different parity than that of the resonance, b) only few particle decay channels are open in medium-heavy nuclei at the energy of the giant dipole resonance, c) the weakness of the electromagnetic coupling.

Beyond mean field, giant resonances can also relax by coupling to progressively more complicated states lying at the same excitation energy. The first in this hierarchy of states are 2p–2h excitations. The most effective among them are those containing an uncorrelated particle–hole excitation and a collective surface vibration. The coupling to these "doorway states" leads to a breaking of the giant resonance strength. This is the phenomenon of collisional damping, and is essentially connected with the coupling of the giant modes to the quantal fluctuations of the nuclear surface (cf. Fig. 1.8). The associated spreading width provides an overall account of the experimental findings. Coupling to more complicated states, and eventually to the totally mixed states describing the compound nucleus, gives rise to a smoothly varying strength function. The associated damping width does not differ substantially from the spreading width arising from the doorway coupling (cf. Sect. 4.2.2).

Collisional damping of giant resonances is found to be be essentially independent of temperature. This in spite of the fact that single-particle damping displays a linear dependence with T. Quantum interference between the different amplitudes involving the decay of a particle or of a hole, essentially cancels out this dependence (cf. Ch. 9). Furthermore, the coupling to the fully mixed states describing the compound nucleus is also independent of the excitation energy. This is because the increase in the level density is compensated by the complexity of the wave function describing the compound states (cf. Ch. 4).

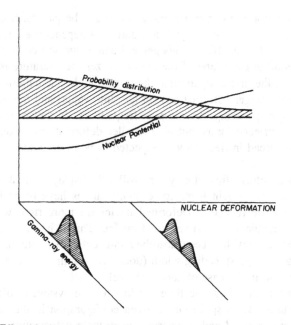

Figure 1.9. Effect of thermal fluctuations on the line shape of giant resonances. The potential energy of the nucleus depends on its deformation, and at finite excitation energy there is a statistical distribution of probabilities for various deformations (after Bertsch and Broglia (1986)).

The study of the collisional damping of nuclear motion is intimately connected with the study of the interweaving of the elementary modes of excitation (namely single-particle vibrations and rotations) in cold nuclei (Bohr and Mottelson (1975)). The corresponding dressed excitations can be viewed as manifestations of the damping phenomena in its incipient form. It is expected that the study of the associated damping phenomena in moderately excited nuclei, as observed in the case of rotational damping in warm nuclei (Ch. 11), will provide important information on the properties of nuclear motion in excited nuclei. This expectation is common with that developed in different fields of physics research, and reflects the idea that it is more difficult that important new results can emerge from the study of systems with a high degree of order and stability, such as crystals. On the other hand, completely chaotic systems, such as turbulent fluids or heated gases, are too formless. The rule seems to be that truly complex things appear at the border between rigid order and randomness.

The contribution of inhomogeneous damping to the lifetime of the giant resonance, is expected to depend on temperature. This is because at finite temperatures, the system will have a larger probability to display deformations which are further removed from the equilibrium deformation (cf. Fig. 1.9). These probabilities are controlled by the Boltzmann factors, and the resulting spread in frequencies will result as a weighting average over all possible quadrupole distortions. Because the corresponding integral can be parametrized in most cases in terms of a Gaussian integral, a square root dependence with temperature will follow (cf. Sect. 10.6). This general behaviour will be "qualified" by whether the equilibrium shape of the system at zero temperature is spherical or deformed. In the first case, thermal, large amplitude fluctuations will make the system probe larger deformations (in absolute value) and this is expected to lead to a large spreading of the dipole frequencies. In the second case, the system will experience larger but also smaller deformations. Consequently, the associated spread in frequency is expected to be more modest than in the first case.

Around each deformation, the system will still display quantal, small amplitude fluctuations. Again here, some qualifications have to be brought in to make the picture realistic. Deformed nuclei are more rigid with respect to quadrupole fluctuations than spherical nuclei. This is because most of the quadrupole collectivity has been absorbed into collective rotations. Consequently, the associated spreading width (doorway coupling), is expected to be less effective than in the case of spherical nuclei.

As the temperature rises, the time it takes for the system to diffuse from a given orientation Ω in space and a given configuration in the (β, γ)-plane (where the parameters β and γ measure quadrupole deformation of the system, cf. Bohr and Mottelson (1975)) to a different (β, γ, Ω)-configuration will decrease. In other words, the time the nucleus conserves memory of a given deformation and orientation, which is connected to the spreading width of multi-quasiparticle states (cf. Sect. 9.1), is expected to decrease with temperature. When these times are shorter than the times associated with the frequency distribution induced by surface coupling, the giant dipole resonance will not be able to fully experience the different shapes. Inhomogeneity in space will be averaged out by inhomogeneity in time. This is the phenomenon of motional narrowing (cf. Sects. 10.7 and 11.3), well known from nuclear magnetic resonance studies in solid state physics (cf. e.g., Slichter (1963)), which can reduce in an important way the rate of increase with temperature of the damping width of a resonance (cf. e.g., Dattagupta (1987)).

Summing up, due to the fact that collisional damping does not depend on the intrinsic excitation energy of the system, that inhomogeneous damping displays a weak dependence with T, and that motional narrowing can make this dependence even weaker, giant resonances are expected to exist as well defined excitations up to temperatures as high as few MeV.

A different question is, whether the eventual γ-decay of a giant resonance can be observed at all values of T. This is because the time it takes for a giant resonance to exist as a thermal excitation in a compound nucleus after the heavy ion fusion process has taken place, may be longer than the time it takes for the compound nucleus to cool through particle emission. Consequently, the γ-decay of the giant dipole resonance, may not be observable, within the limits of sensitivity of the available γ-arrays, at all temperatures. The highest temperature for the observation of the γ-decay of the giant dipole resonance is expected to correspond to that in which the damping width of the giant vibration and the width for particle emission from compound states have similar values (cf. Sect. 6.2).

Before concluding this section it may prove useful to remind the reader of the fact that the use of the concepts of damping and of lifetime for processes in which the energy and the angular momentum of the collective mode do not leave the nucleus, but are redistributed among all its degrees of freedom, leading to the compound nucleus should be used with care. This is because what in fact happens is that the strength of the collective mode is broken by the coupling to more complicated configurations, starting from those denoted "doorway states" (i.e., 2p–2h configurations containing a low lying surface vibration), and including all possible np–nh excitations entering in the compound nucleus wavefunction. To the extent that these more complicated configurations are sharp, each of the resulting mixed states will also be sharp and display infinite lifetime. Consequently, while the coherence existing between the compound states at the time when the collective mode is excited is lost within a time which, in the case of giant dipole resonance is of the order of 10^{-22} sec, each compound state within the range of the FWHM of the GDR resonance is still well defined after 10^{-15}–10^{-16} sec. This is the time needed for the vibrating system to emit a photon. It is through such a process, as well as particle decay, which occurs after $\approx 10^{-18}$ sec, that each compound state carrying a fraction of the giant resonance strength acquires a finite lifetime.

1.5 Low-Lying Vibrations

The lowest excited states of even–even nuclei display, with very few exceptions, quadrupole and octupole character. The associated energies, 1–2 MeV, show an energy dependence $\sim A^{-2/3}$ typical of surface vibrations (cf. Fig. 1.10). They reflect the plastic properties of atomic nuclei. In fact, away from closed shells, the energy of the lowest 2^+ state can become particularly low in energy (cf. Fig. 1.5). In these nuclei, the 2^+ state is the lowest excited member of the so-called ground state rotational band, which can be described in terms of a spheroid rotating along an axis perpendicular to the symmetry axis. These states are excited with large cross sections by Coulomb and by nuclear fields which change slowly in time ($1/\omega \sim 10^{-20}$ sec), as compared with typical times associated with single-particle motion.

In order to excite giant resonances with sizable cross sections, use has to be made of fields which change fast in time, with frequencies of the order of those associated with single-particle motion ($\hbar/\bar{\varepsilon} \approx 3 \times 10^{-23}$ sec, where $\bar{\varepsilon} \approx (2/3)\varepsilon_F \approx 24$ MeV is the average energy of a nucleon, ε_F being the Fermi energy).

Systems which display both elastic and plastic behaviour are well known from other fields of physics. In particular, the natural caoutchouc, that is, the rubber out of which the tires of cars are made. The basic ingredient

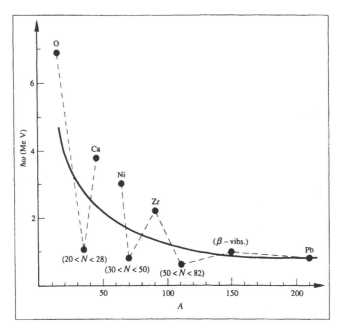

Figure 1.10. Systematics of low-energy quadrupole excitations (from Broglia et al. (1994)).

of caoutchouc is a liquid, known as latex. It is made out of long polymer chains containing only carbon and hydrogen atoms. When this liquid is boiled together with sulphur, which fixes the polymers among them at certain points, it becomes caoutchouc. In this reaction, only one carbon atom out of 200 of them reacts with the sulphur. Consequently, over long distances the atoms of the polymers occupy reasonably fixed positions, as in a solid, while over short distances they are still quite free, as in a liquid. Under a stress, short in time as compared with the diffusion time of the atoms, the liquid part of the material has no time to change place allowing a change in position of the fixed points, and the system reacts elastically, displaying a high degree of buoyancy. When the system is subject to an external field over time of the order or longer than the diffusion time of the polymers, the fixed points have the possibility to change position in space and the caoutchouc deforms plastically (cf. e.g., de Gennes (1994)).

In the nucleus, the rigidity is provided by the single-particle shells. The typical time associated with these fixed points is $1/\omega_0 \approx 7 \times 10^{-23} sec$, where $\hbar\omega_0 \approx 9$–10 MeV is the energy difference between major shells in medium-heavy nuclei. When the atomic nucleus is subject to an external field which changes slowly in time ($\Delta t \gg 1/\omega_0$), the system deforms, and the single-particle levels split in energy. The residual interaction acting among nucleons, in particular the pairing interaction, makes pairs of particles change single-particle orbitals in connection with the crossing of empty and occupied levels, and the system deforms plastically. Evidence of this behaviour is, of course, provided not only by low-lying vibrations and rotations, but also by nuclear fission and exotic decay (cf. e.g., Barranco et al. (1992), Broglia et al. (1993)).

Because the plastic behaviour of the nucleus depends on subtle effects like the detailed occupation of single-particle levels close to the Fermi energy, the superfluidity of the system, the surface tension, etc., low-lying vibrations and rotations are more strongly affected by temperature than giant resonances. The detailed behaviour of low-lying vibrations and rotations as a function of temperature, although being a central chapter in the study of the nuclear structure of hot nuclei, is still a rather unexplored question.

1.6 Rotations

It is well known that nucleons in the atomic nucleus can organize their motion, leading to quadrupole deformed shapes of the average field (intrinsic states), and to rotations of the nucleus as a whole (cf. Fig. 1.5). Discrete rotational bands have been observed extending up to angular momenta higher than 64 units of $\hbar$. On the other hand, the question of how a warm nucleus rotates is still quite open, and the concepts to attack this question are only now being developed. To do so one can try to make use of the obvious analogies that can be made between the damping of the variety of collective motions in finite quantum systems without much avail. This is because, for example, giant resonances have large excitation energies and even when based on the ground state, they are embedded in a background of high density of complicated states to which they couple, eventually acquiring a width. On the other hand, the two states at spin I and spin $I - 2$ connected by a collective quadrupole rotational transition are essentially at the same excitation energy, as measured from the lowest states of the corresponding angular momenta (so called "yrast band"). In a statistical treatment, this fact implies that along the whole rotational band, the density of background states to which the rotational states couple, essentially does not change. Consequently, the eventual damping of rotational motion due to this coupling will be associated with the different behaviour rotational bands display as a function of rotational frequency. In fact, if all the rotational bands behaved equally as a function of the rotational frequency, their coupling would lead to a new set of rotational bands displaying the same sharp transitions as the original set (cf. Sect. 11.4). The intrinsic states, on which the rotational bands are built, react instead in different ways to Coriolis and centrifugal forces. Because of this variety of responses, the rotational decay from a state at spin I will not populate a single state at spin $I - 2$, but rather a whole distribution of them. This spreading of the rotational transitions is the phenomenon of rotational damping (cf. Fig. 1.11). The width of the associated strength function, the rotational damping width, measures the time it takes for the rotating system, to lose coherence at the level of the constituting intrinsic configurations. A new facet associated with the study of

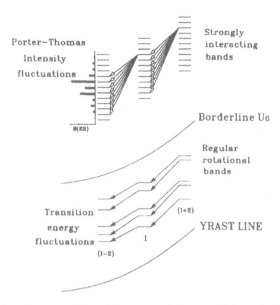

Figure 1.11. Schematic representation of the regime of discrete and damped rotational bands. After T. Døssing et al. (1996).

rotational damping, as compared to that of vibrational motion, is connected with the fact that to properly identify transitions within a band, one needs to make use of coincidence measurements of gamma-rays. In particular, the higher folds (higher number of gamma-rays measured in coincidence) are very important in this quest.

The various patterns expected in the two-dimensional coincidence γ-spectra are schematically shown in Fig. 1.12. The coincidence γ-spectrum of the decay of a single rotational band displays a very simple structure, as shown in Fig. 1.12(b). In particular, there are no events on the diagonal $E_{\gamma_1} = E_{\gamma_2}$, and each event is separated from the neighbour event by an energy which is inversely proportional to the moment of inertia. Making a cut perpendicular to the diagonal (cf. Fig. 1.12(c)) one observes a very regular distribution of peaks. In the case of a number of discrete rotational bands displaying different moments of inertia and alignments, i.e., the amount of angular momentum carried out by specific individual configurations, the same pattern is observed but somewhat more fuzzy (cf. Fig. 1.12(e)). The cut perpendicular to the diagonal still leads to a series of regular peaks and valleys (cf. Fig. 1.12(f)). Already at about 1 MeV above the yrast line, the density of levels is sufficiently high for the basis rotational bands to mix strongly. In other words, any state of a strongly rotating hot compound nucleus will be a linear combination of unperturbed rotational bands displaying different alignments and moments of inertia, and the gamma decay will reflect this admixture. The $E2$-decay out of a state at spin I will not go to a unique state, as it does for bands close

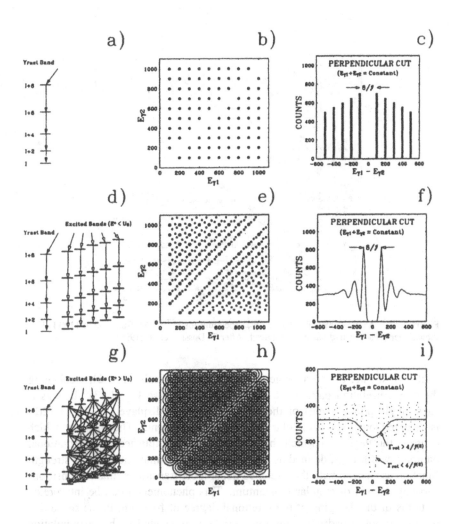

Figure 1.12. This schematic figure shows the relation between the different decay modes in form of level schemes in the three left panels, the corresponding patterns expected in the two-dimensional (2D) $E_{\gamma_1} \cdot E_{\gamma_2}$ spectra in the middle panels and finally in the right side, the corresponding one-dimensional (1D) perpendicular cuts performed for constant $E_{\gamma_1} + E_{\gamma_2}$ on the 2D spectra shown in the middle panels. Panels (a) and (b) show how one regular rotational band with a constant moment of inertia generates a grid pattern in $E_{\gamma_1} \cdot E_{\gamma_2}$ spectra, with the diagonal missing. Panels (d) and (e) display the pattern of ridges generated by an ensemble of bands with variations in alignment and moment of inertia, the ridges becoming wider as one moves away from the diagonal $E_{\gamma_1} = E_{\gamma_2}$. Panels (g) and (h) describe the situation of rotational damping. Here, the average transition energies form a grid pattern, which is then smeared out in both directions by a width Γ_{rot}. This results in a smooth spectrum with a valley in the middle. Cuts in the direction perpendicular to the diagonal are shown in panels (c), (f) and (i) for the three different situations (from B. Herskind et al. (1992)).

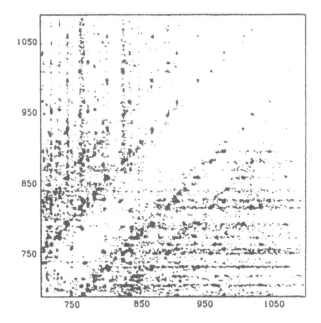

Figure 1.13. An experimental 2D $E_{\gamma_1} \cdot E_{\gamma_2}$ spectrum for ^{167,168}Yb from 700 to 1100 keV is shown after a background has been removed. After T. Døssing et al. (1996).

to the yrast line, but to a range of final states with spin $(I - 2)$, weakening the γ–γ correlation pattern (cf. Fig. 1.12(g–i); cf. also Fig. 1.13). It is as if each single point (event) of the discrete spectrum displayed in Fig. 1.12(b) is smeared out over a circular area around the event. The quantity which controls the degree to which this smearing is accomplished is the rotational damping width Γ_{rot}, defined as the full width at half maximum of the strength function associated with the stretched quadrupole transitions, that is transitions carrying 2 units of angular momentum. This phenomenon can be interpreted in terms of the damping of the rotational degree of freedom, in the sense that the quadrupole–quadrupole response function associated with the population of a mixed band state displays a finite lifetime $\hbar/\Gamma_{rot}$. In this damped region, the γ-decay from angular momentum I to $I - 2$ will spread out over many possible transitions. As a consequence, the decay through consecutive angular momenta, $I \rightarrow (I - 2) \rightarrow (I - 4)$ will populate many decay paths. In other words, for each compound nucleus formed in an experiment, the cascade of $E2$ γ-rays which eventually will cool the system, will find many transitions through which to proceed in the region where bands mix strongly (damped region) and only few in the region of discrete bands.

An example of an experimental γ-coincidence spectrum is shown in Fig. 1.13. It displays a system of ridges spaced 60 keV apart, and running parallel to the main diagonal, confirming, to a large extent, the schematic picture of Fig. 1.12(e,f). Along the diagonal there exists a rather featureless domain. This is in fact a valley, which represents a 10% depression relative to the surroundings, in correspondence with Fig. 1.12(h,i). Such a coincidence γ-spectrum carries information on the properties of rotational motion as a function of temperature along the region where the decay flow goes through, a subject which is taken up again in Chapter 11.

Part 1

ZERO TEMPERATURE

2

Giant Vibrations

A variety of resonant states, at energies larger than the nucleon separation energy (8–10 MeV) have been observed throughout the mass table. They are known as giant resonances or giant vibrations, in keeping with the fact that they display large cross sections, close to the maximum allowed by sum rule arguments, implying that all nucleons participate in the vibration.

2.1 The Variety of Modes

Giant resonances have been extensively studied during the last decade. These modes can be classified making use of the multipolarity L, spin S and isospin T quantum numbers (cf. Fig. 2.1) according to:

a) Electric ($\Delta S = 0$) isoscalar ($\Delta T = 0$) vibrations where protons and neutrons oscillate in phase according to a multipole pattern, for example quadrupole, as shown in Fig. 2.2, and the electric isovector ($\Delta T = 1$) vibrations where protons oscillate against neutrons (cf., e.g., Fig. 1.2). For the same multipolarity, the isovector modes lay at a higher excitation energy than the isoscalar vibrations, due to the extra energy required to separate neutrons from protons,

b) Magnetic or spin-flip ($\Delta S = 1$) isoscalar ($\Delta T = 0$) modes associated with the motion of nucleons with spin up which oscillate against nucleons with spin down. In the corresponding isovector modes ($\Delta S = 1$ and $\Delta T = 1$), protons with spin up oscillate against neutron with spin down and vice versa (cf. Fig. 2.3).

At the microscopic level, giant resonances can be viewed in terms of correlated particle–hole excitations. In the case of the giant dipole resonances, and making use, for simplicity, of a harmonic oscillator potential to describe the single-particle motion, the most important excitations are those corresponding to $\Delta N = 1$ transitions, that is particle–hole excitations which change by 1 the oscillator principal quantum number N. This in keeping with the fact that the dipole mode has negative parity and that the parity of the single-particle orbital

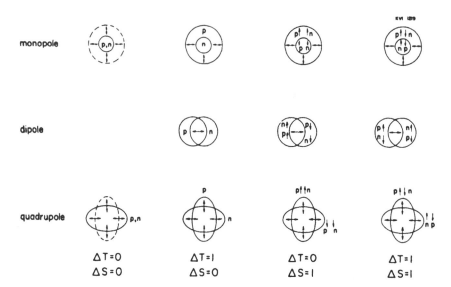

monopole

dipole

quadrupole

$\Delta T = 0$
$\Delta S = 0$

$\Delta T = 1$
$\Delta S = 0$

$\Delta T = 0$
$\Delta S = 1$

$\Delta T = 1$
$\Delta S = 1$

Figure 2.1. Schematic representation of the nuclear collective modes with multipolarity L = 0 (monopole) L = 1 (dipole) and L = 2 quadrupole associated with vibrations of all nucleons in phase ($\Delta T = 0$, isoscalar modes) of protons in phase opposition with neutrons ($\Delta T = 1$, isovector modes). The spin modes are also illustrated. The ones in the third column correspond to vibrations of protons and neutrons with spin up in phase opposition with those with spin down, classified as $\Delta T = 0$, $\Delta S = 1$ modes, while the ones in the fourth column correspond to vibrations of protons with spin up and neutrons with spin down in phase opposition with those of protons with spin down and neutron with spin up, classified as $\Delta T = 1$, $\Delta S = 1$ modes. (After Van der Woude (1987)).

belonging to a shell N of the harmonic oscillator is $(-1)^N$. In the case of quadrupole resonances, all excitations which change by an even number the principal quantum number are allowed by parity considerations (see Fig. 2.4). Among them, $\Delta N = 2$ excitations control the energy centroid of the resonance, while $\Delta N = 0$ excitations give rise to low-lying surface collective vibrations.[1]

A number of the giant vibrations schematically shown in Fig. 2.1 have been observed making use of a variety of experimental probes (cf., e.g., Van der Woude (1987)). They are: i) the isoscalar monopole, quadrupole and octupole giant resonances, ii) the isovector monopole, dipole and quadrupole giant resonances, iii) the Isobaric Analog and Gamow–Teller giant resonances. In what follows we shall discuss examples of both isovector and isoscalar modes.

---

[1] Excitations of $\Delta N = 2$ type in the case of the quadrupole modes are associated with the elastic, short time response of the system, while $\Delta N = 0$ excitations are connected with the plastic response of the average potential to slow changing external fields (cf. Sects. 1.4 and 1.5).

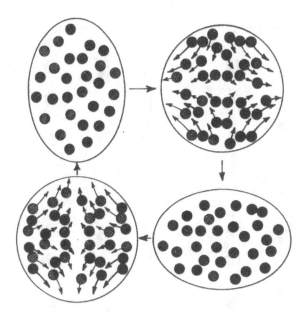

Figure 2.2. Pictorial view of the isoscalar giant quadrupole vibration. Protons (lighter circles) and neutrons (heavier circles) are all excited by a projectile that interacts with them through the strong interaction. (after Bertsch (1983)).

2.2 The Giant Dipole Resonance

The first experimental evidence for the excitation of a simple vibratory motion in nuclei was obtained in 1937 in measurements of the radioactivity produced in a variety of targets by a source of 14 MeV photons. In some of the targets used, it was found that the photoabsorption cross section displayed a resonant behaviour. Improved data obtained with bremsstrahlung photon beams from electron accelerators (Berman and Fultz (1975)), confirmed the resonant behaviour of the photoabsorption phenomenon, and provided information concerning the parameters of the associated reaction cross section. Because the resonance energy ($\hbar\omega_D \approx 15$ MeV) corresponds to a wavelength of the photon ($\lambda \approx 100$ fm) which is much larger than the nuclear radius (R = 5–7 fm), the nucleus is exposed, when shined with a beam of photons of the corresponding energy, to a uniform time-dependent electric field, which causes all protons to move in the same direction. Since in the photoabsorption process, the center of mass of the nucleus is at rest or in uniform motion, the neutrons have to move in opposite direction to that in which the protons move (cf. Fig. 1.2). The strong force acting among nucleons provides the restoring force of this vibrational mode (cf. Ch. 3).

Progress in the study of the giant dipole vibrations was made when monoenergetic photon beams became available. The techniques employed to produce

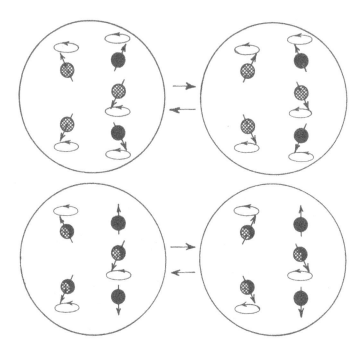

Figure 2.3.   Schematic illustration of spin vibrations. In the giant magnetic-dipole resonance (top), a proton and a neutron are tipped in opposite directions, that is their spin axes process 180° out of phase with respect to each other. As a result of the correlated precession of a fraction of the nucleons the nucleus as a whole acquires a net spin and a net magnetic moment. In the bottom part an illustration of the Gamow–Teller resonance is given. A proton is converted into a neutron and its spin is tipped from the original orientation. (After Bertsch (1983)).

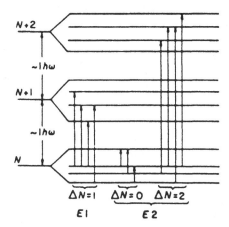

Figure 2.4.   Schematic representation of E1 and E2 single-particle transitions between shell model states.

such beams with energies up to approximately 50 MeV are mainly two: a) positron production followed by annihilation in a secondary target, b) selection of bremsstrahlung radiation produced in the slowing down of electrons by the protons of the target nucleus, leading to particular kinetic energies of the outgoing electrons (tagged photon methods). Typical experiments for the measurement of photon reaction cross section use large volume NaI(Tl) scintillators to detect the incident integrated number of photons and efficient neutron detectors (or charge particle detectors in the case of light nuclei), to reveal the occurrence of a nuclear reaction. Making use of these techniques, systematic photoabsorption studies were carried out throughout the mass table. The common feature observed is that the resonance can be well fitted in terms of one or two Lorentzian curves (cf., e.g., Figs. 1.4 and 1.6). The magnitude of the total cross section is proportional to the area subtended by these curves. Comparing such quantity to the theoretical limit known as the dipole oscillator strength or energy weighted sum rule (cf. Eq. (3.40)), one can estimate the number of nucleons participating in the excitation. From this comparison it results that essentially all nucleons participate in the vibration, that is, the dipole vibration exhausts essentially 100% of the oscillator strength (cf. Fig. 2.5(a)). This is the reason why the dipole resonance is called giant.

The centroid energy of the Lorentzian curve which fits the resonance is, in the case of medium-heavy mass nuclei, well reproduced by the relation (cf. Fig. 2.5(b)),

$$\hbar\omega_D \approx \frac{79}{A^{1/3}} \ \text{MeV}. \tag{2.1}$$

In the case of quadrupole deformed nuclei, the absorption resonance cross section is generally reproduced by two Lorentzian functions, whose centroids and relative intensities can be used to infer the size and the type of nuclear deformation (cf. Figs. 1.6 and 1.8). In fact, if one denotes by δ the ratio between the energy centroid of the higher and lower components of the resonance, it is possible to express the quadrupole deformation parameter β according to

$$\beta = \sqrt{\frac{4\pi}{5}} \left(\frac{\delta - 1.0}{0.5\delta + 0.87} \right). \tag{2.2}$$

This relation was first deduced by Danos (1958) starting from Bohr's parametrization of the nuclear radius (Bohr and Mottelsson (1975)) and leads to results for β which are in overall agreement with those obtained from other sources, in particular from the energies and quadrupole transition probabilities associated with members of the ground state rotational band of the corresponding nuclei.

The type of quadrupole deformation displayed by the system can be deduced from the ratio of the areas subtended by the higher and lower components of the giant dipole resonance. For an oblate shape this ratio is 0.5, while for a prolate shape it is 2, as shown in Fig. 2.5(e). The properties characterizing the nuclear quadrupole deformation, as obtained from photoabsorption measurements, are displayed in Figure 2.5(d) as a function of the mass number. These data provide

GROUND STATE GDR

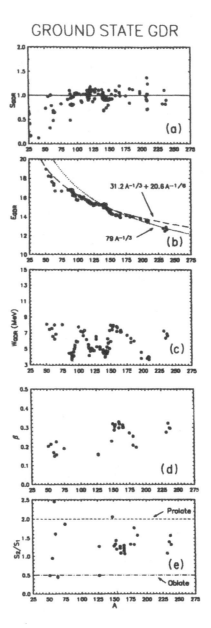

Figure 2.5. Observed properties describing the giant dipole resonance based on the ground state as a function of the nuclear mass number A. (a) Strength in units of the energy weighted sum rule, (b) centroid energy, (c) width, (d) quadrupole deformation parameter β as obtained from the giant dipole resonance lineshape (cf. Eq. (2.2)), (e) ratio of the strength in the high-energy component denoted with S_2 with that of the low energy component S_1. (Adapted from Gaardhøje (1992).)

experimental evidence of the coupling of the giant dipole resonance to the quadrupole deformation of the nuclear surface. Based on angular momentum and parity selection rules, this coupling is expected to be the most important in breaking the resonance strength. Other types of deformations, such as octupole can also couple to the GDR leading to spin and parity 1^-, but only in second order. Monopole distortion of the mean field can also couple to the giant dipole vibration, as quadrupole deformations do, in first order. On the other hand, one expects this coupling to lead to smaller dispersions of the dipole frequency than those associated to the quadrupole channel because of two reasons: a)only dynamical effects are operative in the monopole channel, b)low-lying modes are, as a rule, absent in the monopole channel.

The damping width of the giant dipole resonance defined as the full width at half maximum (FWHM) of the photoabsorption reaction cross section, is displayed in Figure 2.5(c) as a function of mass number. In keeping with the discussion above, quadrupole deformed nuclei have a width (peaks in Fig. 2.5(c), cf. also Fig. 4.10) which is larger than typical widths associated with the giant dipole resonance in spherical nuclei.

The special role that quadrupole distortions of the nuclear surface play in the line shape of the giant dipole resonance allows also to extract, from the width of the giant dipole resonance in spherical nuclei, the effective quadrupole deformation parameter $\langle \beta^2 \rangle^{1/2}$ (cf. Fig. 2.5(d)). The corresponding values agree well with those obtained through other spectroscopic methods, in particular inelastic scattering or Coulomb excitation connecting the ground state with the low-lying 2^+ state. This fact will be exploited in Chapters 5,6 and 10 to study the evolution of the nuclear shape and of surface fluctuations as a function of temperature and rotational frequency.

2.3   The Monopole and Quadrupole Vibrations

While photons are an accurate probe of nuclear structure, they excite essentially only the giant dipole resonance. This is because of the long wavelength of the photon as compared to the dimensions of the system. To study non-dipole modes, use of probes different from photons, like hadrons, have to be made.

The first giant resonance discovered in hadron scattering was the giant isoscalar quadrupole vibration (Bertrand (1976)). This vibration, unlike the dipole, is a shape vibration (cf. Fig. 2.2). A nucleus vibrating in the quadrupole mode is distorted from a spherical shape to an ellipsoidal shape and moves back through a spherical shape to an ellipsoidal shape of another orientation. From this picture it emerges that, in order to excite shape vibrations, it is necessary, not only to impart a given energy to the nucleus, but also to distribute this energy in such a way that the nucleons are set into motion in different directions. This can be accomplished, for example, by inelastic scattering of alpha particles.

Typical spectra of the kinetic energy in the laboratory system of scattered alpha particles from several medium-heavy targets are shown in Figure 2.6. In these spectra, the strong peaks associated with the excitation of the low-lying states of the target nucleus have been scaled down, to emphasize the presence of a broad bump in the unbound region ($E_x > 10$ MeV). The main evidence that this bump might be associated with the inelastic excitation of a quadrupole vibration, was obtained from the angular distribution of the α-particle differential cross section (cf. Fig. 2.7). In fact, each multipolarity leads to a characteristic angular distribution. If the wavelength of the relative motion is short compared to the dimensions of the target nucleus, the angular distribution of the α-particle peaks at an angle determined by angular momentum conservation. Denoting by $\mathbf{k}_i$ and $\mathbf{k}_f$ the wavenumber vectors of relative motion in the entrance and exit channel respectively, and with R the nuclear radius, the transferred angular momentum can be written as

$$\Delta l = |\mathbf{k}_i - \mathbf{k}_f|R = qR, \tag{2.3}$$

the quantity $\hbar q$ being the linear momentum transferred to the target nucleus. Geometrical considerations can be used to obtain the modulus square of $\Delta l = L$, where L is the multipolarity of the excitation, namely,

$$(\Delta l)^2 = R^2 q^2 = (k_i^2 + k_f^2 - 2k_i k_f \cos \theta)R^2. \tag{2.4}$$

For small values of θ, $k_i \approx k_f \approx (2\pi/\lambda)$ and $(1 - \cos\theta) = 2\sin^2\theta/2 \approx \theta^2/2$ leading to

$$\theta = \frac{\lambda}{2\pi R}L. \tag{2.5}$$

In the case of the excitation of a state at ≈ 11 MeV in ^{208}Pb by a beam of 100 MeV alpha-particles, Eq. (2.5) gives, for $L = 2$, a scattering angle $\theta \approx 4\text{--}5°$, as experimentally observed.

In contrast, one can see from Figure 2.7 that the inelastic scattering of α-particles exciting the isoscalar giant monopole resonance (lying at an energy of 13.7 MeV in ^{208}Pb) leads to an angular distribution peaking at 0°. From the same figure it is seen that at 4°, where the angular distribution of the giant quadrupole resonance has a maximum, the cross section for the excitation of the giant monopole resonance has a minimum.

The excitation of the giant monopole and quadrupole resonances has been also studied in inelastic processes induced by heavy ions (for example ^{17}O). In this case, the excitation cross section was found to be larger than for alpha-particles, although the angular distributions did not have the characteristic patterns shown in Figure 2.7 (Bertrand et al. (1988)).

The importance of studying the giant monopole resonance, the so-called breathing mode, is because this is one of the few ways in which one can obtain direct information on the nuclear matter compressibility modulus K_{nm}. This quantity is an important ingredient of the nuclear matter equation of state, and as such plays a central role in the discussion of a variety of phenomena

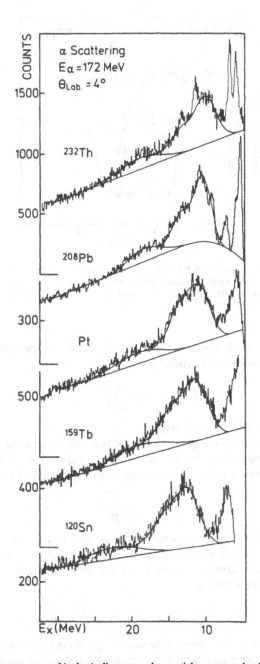

Figure 2.6. Energy spectra of inelastically scattered α-particles measured at bombarding energy $E_\alpha = 172$ MeV and at a scattering angle of 4° for a number of heavy nuclei. A possible decomposition into a continuum and various resonant curves is indicated. (Adapted from Morsch et al. (1982).)

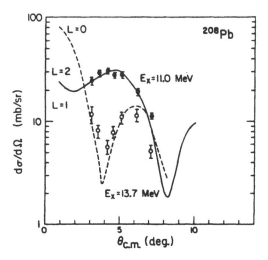

Figure 2.7. Angular distribution for inelastic α scattering on ^{208}Pb. The solid line is the theoretical distribution for exciting a quadrupole vibration $E_x = 11$ MeV L = 2 and the dashed line is the theoretical distribution for the monopole $E_x = 13.7$ MeV, L = 0. (Adapted from Youngblood et al. (1977).)

like in supernovae explosions and in relativistic heavy ion collisions. The compressibility modulus is defined as

$$K_{nm} = 9\rho_0^2 \frac{d^2 E/A}{d\rho^2}|_{\rho_0}, \qquad (2.6)$$

where ρ_0 is the equilibrium density.

The energy $\hbar\omega_M$ of the giant monopole resonance extracted from the experimental findings, is related to the effective compression modulus K_A for the nucleus of mass number A, according to an expression of the type

$$\hbar\omega_M \sim \sqrt{\frac{K_A}{R^2}}, \qquad (2.7)$$

the proportionality coefficient being model dependent (cf., e.g., Bertsch and Broglia (1994), Blaizot et al. (1995)). The quantity K_{nm} is then obtained, using for K_A an expansion inspired by the semi-empirical mass formula (Weizsäcker (1935)), that is,

$$K_A = K_{vol} + K_{surf}A^{-1/3} + K_{sym}\frac{(N-Z)^2}{A^2} + \cdots \qquad (2.8)$$

The determination of the various parameters K_i is done by a fit to the empirical data, and K_{vol} is identified with K_{nm}. However, because of the limited set of data existing, equally "good fits" can be obtained with the volume term taking values in the range 100–400 MeV (Pearson (1991), Shlomo and Youngblood (1993), Blaizot et al. (1995)). Values of the order of 210 MeV are obtained making use of the microscopic formalism described in Chaps. 3 and 4.

2.4 The Decay of Giant Resonances

Typical values of the centroid and of the FWHM of giant resonances are 10–15 MeV and 3–5 MeV, respectively. Consequently, giant vibrations go only through few periods of oscillation before they relax. There are a variety of mechanisms which can damp giant vibrations. They can first be characterized by whether the energy of the vibration escapes the system, or whether it is redistributed into other degrees of freedom within the system. In the first class of damping mechanism one finds γ-emission and particle-decay, characterized by a width Γ_γ and $\Gamma^\uparrow$, respectively. Widths due to photon emission are very small as compared to the typical FWHM of the resonant curve ($\Gamma_\gamma/\Gamma \approx 10^{-4}$). Particle emission, on the other hand, is not negligible measured in the same scale ($\Gamma^\uparrow/\Gamma \approx 10^{-1}$).

Particle decay from giant resonances can be studied in experiments where the inelastic scattered particles exciting the giant resonance are measured in coincidence with particles emitted in the decay of the resonance. The energy interval spanned by the spectrum of these two types of particles is very different. Consequently, two different kinds of detectors are usually employed in the measurements. Because of the large Coulomb barrier associated with charged particles in medium-heavy nuclei, particle-emission from giant resonances consists mainly of neutrons. In keeping with this fact, a large fraction of the experimental work is based on neutron spectra and multiplicity.

The neutron detectors employed in the measurements are either plastic or liquid scintillators. The energy of the neutrons is measured by time of flight techniques. In the case in which liquid scintillators are used, a pulse shape analysis can also be made. This additional information allows to discriminate, in the energy region of the neutrons of interest, the contributions arising from room background, as well as gamma-rays due to activation. The neutron detectors are also shielded to reduce the counting of the neutrons originating from other sources besides the target, such as the Faraday cup and the beam guiding system.

An alternative method employed to study the neutron decay from giant resonances is that of measuring the excitation energy of the residue of mass number $(A-1)$, after neutron emission from nucleus A has taken place (Bracco et al. (1989)). This can be accomplished by measuring the summed gamma-ray energy making use of a 4π-detector array. Because the times involved in the variety of decay processes are very short ($\approx 10^{-21}$–10^{-20} sec), it is not possible to know wether particle decay following the excitation of a giant vibration has taken place before or after this mode has been damped out (cf. Fig. 2.8). The separation between direct- and compound-decay is eventually obtained, through model dependent calculations.

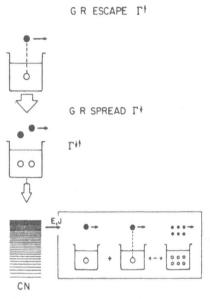

Figure 2.8. Schematic illustration of the neutron decay of a giant resonance state. The giant resonance is represented with filled and empty circles connected by a dashed line (upper part). In the middle part the giant resonance has coupled to two-particles two-holes states and from there on can decay by either particle emission or by coupling to more complicated states. The final stage of the damping is the coupling to the compound nucleus (lower part). When $\Gamma^\uparrow$ is different from zero, the decay of the compound nucleus, into which the resonance has relaxed, contains a contribution from the GR direct decay.

2.5 Direct and Compound Particle-Decay

The statistical component of particle decay arising from the evaporation of the compound nucleus at excitation energies in the energy region corresponding to the centroid of a giant resonance, can be inferred through a comparison between the observed particle spectra, and the spectra calculated employing a statistical model of nuclear decay. The calculations are, as a rule, performed making use of the Hauser–Feshbach model (Hauser and Feshbach (1952)), assuming that all the states of the nucleus A with excitation energy E_i and spin and parity J_i^π, are equally populated. In this case, the probability that the system decays through the emission of a particle p with kinetic energy ϵ_p, leaving the residual nucleus $(A-1)$ in a state E_f with spin and parity J_f^π is given by

$$P_p(E_f)d\epsilon_p = \rho_f(E_f, J_f^\pi)\Sigma_{l_<}^{l_>}\tau_l^p(\epsilon_p)d\epsilon_p, \qquad (2.9)$$

where $l_< = |J_f - J_i - s_p|$ and $l_> = J_f + J_i + s_p$. In the above equation, $E_i = E_f + \epsilon_p - B_p$, B_p being the particle separation energy. The quantity $\rho_f(E_f, J_f^\pi)$ is the level density, in the final nucleus, of states with spin and parity J_f^π at excitation energy E_f, while s_p and l are the spin and the orbital angular momentum of the emitted particle. The quantities $\tau_l^p(\epsilon_p)$ are the transmission coefficients for scattering of particle p with energy ϵ_p and angular momentum l off the nucleus $(A - 1)$. They are obtained from optical model analyses of elastic scattering data.

Different sets of optical potential parameters are currently available leading to stable values of the decay branch (cf., e.g., Rapaport et al. (1979)). In addition, in the case of neutron emission at rather low kinetic energies, it is possible to obtain the transmission coefficients from the measured neutron resonances. For the level density, the actual level scheme is used for the lower excitation part of the spectrum and a level density formula with empirically determined parameters, for the higher excitation energies.

In the cases where the observed branching ratios or experimental spectra cannot be accounted for by the statistical model calculations, the presence of a direct decay component has been invoked, making use of unitarity arguments to provide the interconnection between the two contributions. Following general theoretical arguments (Dias et al. (1986)) and assuming a two-step model, the expression of the energy averaged cross section for the decay of a giant resonance of a system A into the single hole state i of the system $(A - 1)$, is given by

$$\sigma_{GR \to 0} = \sigma_{0 \to GR} \left((1 - \mu_1) \frac{\tau_i^D}{\Sigma_i \tau_i^D} + \mu_1 \frac{\tau_i^C + \mu_1 \tau_i^D}{\Sigma_j \tau_j^C + \mu_1 \Sigma_i \tau_i^D} \right), \qquad (2.10)$$

where $\sigma_{0 \to GR}$ is the cross section associated with the excitation of the giant resonance, starting from the ground state. The label j runs over all the states, including hole states i of the final system, that can be populated from the compound nucleus. The quantities

$$\tau_i^C = 2\pi (\Gamma_i)^C \rho, \qquad (2.11)$$

and

$$\tau_i^D = 2\pi \Gamma_i^\uparrow \rho^D, \qquad (2.12)$$

represent the compound nucleus particle decay and the direct decay transmission coefficients, respectively. In the above expressions $(\Gamma_i)^C$ denotes the compound width, ρ the total density of nuclear states, while $\Gamma_i^\uparrow$ is the escaping width associated with the population of the final state i. The quantity ρ^D is the density of giant resonance states, which is of the order of 1 MeV^{-1}, in keeping with the fact that the FWHM of giant resonances is of the order of few MeV. The mixing parameter μ_1 measures the coupling of the giant resonance to the compound nucleus. It is related to the giant resonance spreading width $\Gamma^\downarrow$ according to

$$\mu_1 = \frac{\Gamma^\downarrow}{\Gamma} = \frac{(\Gamma - \Sigma_i \Gamma_i^\uparrow)}{\Gamma}. \qquad (2.13)$$

The cross section relation given in Eq. (2.10) was obtained applying the multistep compound nucleus formation model (Feshbach et al. (1980)), to the case

Table 2.1. Empirical values for the branching ratios of the neutron decay from the giant monopole resonance in ^{208}Pb. The data in column 2 are from Brandenburg et al. (1989) and those in column 3 from Bracco et al. (1988).

Level	exp. (1)	exp. (2)
$p_{1/2} + i_{13/2}$ (gs + 1.633 MeV)	0.03 ± 0.014	0.046 ± 0.012
$f_{5/2}$ (0.570 MeV)	< 0.014	0.023 ± 0.005
$p_{3/2}$ (0.898 MeV)	0.03 ± 0.016	0.016 ± 0.003
$f_{7/2}$ (2.339 MeV)	0.056 ± 0.012	0.054 ± 0.013

of a two-step process. Figure 2.8 illustrates schematically the content of this model.

The first term of Eq. (2.10) describes the direct decay of the giant resonance into the final channels (cf. Fig. 2.8(a)). Because of the mixing of the giant resonance with the compound nucleus, this term contains a depletion factor $(1 - \mu_1)$. The second term corresponds to the statistical contribution to the cross section. The configurations that are usually taken into account in the Hauser–Feshbach model of the compound nucleus are incoherent np–nh excitations (Fig. 2.8(c)). In the excitation energy region of giant resonances, the description of the compound nucleus should also include coherent particle–hole configurations as an additional class of states of the compound system. This is accomplished in the present model by adding the term $\mu_1 \tau_i^D$ to the usual Hauser–Feshbach transmission coefficient. This term represents the decay of the compound nucleus through the giant resonance which has been thermally excited.

The results obtained applying the two-step model discussed above to the analysis of two sets of data associated with the decay of the giant monopole resonance of ^{208}Pb are reported in Table 2.1. In the case of experiment (1), inelastic scattering of alpha-particles was used and neutrons were detected in neutron scintillator detectors. Experiment (2) used inelastic scattering of ^{17}O and the excitation of the $(A - 1)$ nucleus after neutron emission was obtained through the measurement of its γ-decay. To eliminate background contributions, due to the decay of states in the continuum with unknown structure, two different assumptions were made. In the case of experiment (1) the background underlying the giant monopole resonance bump was assumed to be equal to the cross section in the same energy region at larger angles. Conversely, in the case of experiment (2) the background was taken equal to the cross section measured at the same angle but at larger excitation energy. Because only prompt coincidences were measured, the decay to the ground state and to the long lived isomeric state $i_{13/2}$ could not be distinguished as they both correspond to the absence of gamma-rays following neutron emission.

Recently, the formalism described above has been extended (cf. Teruya and Dias (1994)) to treat particle decay from lighter nuclei where protons and α-particles, in addition to neutrons, are emitted.

2.6 Gamma-Decay of Giant Resonances

The basic property of photon decay is its extreme sensitivity to the multipolarity of the giant resonance. This can be seen from Figure 2.9, where the ground-state gamma width $\Gamma_{\gamma 0}$ expected for a sharp state exhausting 100% of the energy weighted sum rule as a function of multipolarity and energy relative to that for E1 photons is shown. As expected, a large decrease of the decay width at a given excitation energy for increasing electric multipolarity L of the emitted radiation is observed.

This property is quite useful, particularly when giant resonances are populated in inelastic processes induced by heavy ions. In fact, contrary to electron, proton and alpha inelastic scattering, the angular distribution associated with heavy ion inelastic scattering collisions are, as a rule, rather featureless and independent of the angular momentum transferred in the process. This is because the very short wavelength associated with the relative motion of the interacting ions, leads to an essentially classical motion where interference effects are absent. On the other hand, making use of heavy ion reactions at bombarding energies of the order of E_{GR}/u (i.e., at bombarding energies per nucleon, of the order of the energy of the resonance) and higher, it is possible to obtain large cross sections for the excitation of giant resonances, mainly through Coulomb excitation processes. Also, a much more favorable resonance to background ratio (cf. Fig. 2.10) than that obtained in the case of alpha scattering, in keeping with the fact that under the influence of such fast changing fields, low-lying vibrations are little excited (cf. Sect. 1.5). Coincidence experiments benefit from these characteristics of heavy ion processes. To be noted, however, is that Coulomb excitation displays a strong dependence with the multipolarity of the process.

2.6.1 Decay to low-lying states

Photon decay from giant resonances to low-lying excited states is also a potential source of information. These data, like the ground state decay, display strong multipole selectivity but are not limited to $L = 1$ and $L = 2$ resonances. As in the case of ground state γ-transition, decays to excited states are dominated by E1-radiation. Thus, for example, the observation of transitions from the region of giant resonances in the nucleus ^{208}Pb to a low-lying 5^- state, implies the presence of high-lying states with spin 4^+ and 6^+. In general, measurements of γ-decay in coincidence with inelastically scattered particles can help in the determination of the multipolarity of the excited giant modes.

The decay of giant resonances to low-lying states can also provide information about their coupling to low-frequency collective surface vibrations (cf., e.g., Bortignon and Broglia (1981)). For example, the decay of the GQR of ^{208}Pb to the lowest 3^- state of this nucleus, provides a quantitative measure of the role played by 2p-2h states containing an uncorrelated particle–hole

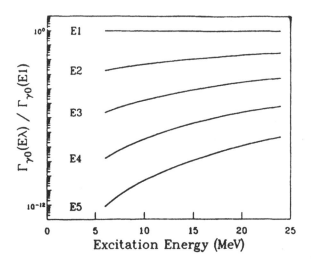

Figure 2.9. Ground-state gamma widths of hypothetical sharp states fully exhausting the appropriate isovector or isoscalar energy weighted sum rule as a function of the excitation energy of the state, relative to the E1 width. (Adapted from Beene et al. (1990).)

coupled to angular momentum 1^- and the octupole surface vibrational mode. It also provides information on the isovector character of the resonance. In fact, in keeping with the fact that low-lying surface vibrations in nuclei along the stability valley are isoscalar, the isoscalar giant quadrupole resonance should display a weak γ-branch to the low-lying 3^- state, much weaker than that associated with the decay of the isovector GQR (cf. Bortignon et al. (1984a), Speth et al. (1985)).

2.6.2 The ground-state decay

A number of measurements of the γ-decay of the nucleus ^{208}Pb from the energy region of giant resonances have been made after exciting the system in inelastic scattering processes induced by a $N \approx Z$ projectile (mainly ^{17}O). Rather high bombarding energies, in the range of 20 to 100 MeV/u, have been used in these experiments (Beene et al. (1989) and (1990)). The cross section for excitation of giant resonances was found to increase with bombarding energy, as seen from Figure 2.10. In addition to the difference in yield with energy, it is apparent from the figure that the giant resonance bump is shifted to higher energies increasing the projectile bombarding energy from 22 to 84 MeV/u. This effect is connected with the fact that the Coulomb excitation of the isovector giant dipole resonance, a mode which essentially is not excited through the hadronic field associated with $N \approx Z$ projectiles, gives the largest contribution to the giant resonance region at higher bombarding energies, whereas the isoscalar quadrupole resonance, excited mainly through

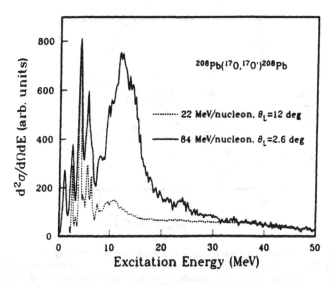

Figure 2.10. Spectra of inelastic scattered ^{17}O from ^{208}Pb for two different bombarding energies, 84 (full drawn line) and 22 MeV/u (dotted line). The spectra are normalized in the unstructured continuum near 40 MeV. (Adapted from Beene et al. (1990).)

the nuclear interaction, provides the largest signal at low bombarding energies. Measurements of the angular correlation of the γ-ray transitions from the giant resonance region to the ground state supports this conclusion (cf. Figs. 2.11 and 2.12). In fact, the angular distribution associated with the 84 MeV/u reaction is well reproduced assuming pure E1-deexcitation. However, to describe the corresponding data of the 22 MeV/u reaction, an E2+E1-mixture is needed. At 22 MeV/u the cross section associated with the giant dipole resonance is small. The largest contribution to the giant resonance bump is due in this case to the nuclear excitation of the isoscalar giant quadrupole and monopole resonances.

The distribution in energy of the ground-state photon decay cross section has been described making use of multistep theory of nuclear reactions already used in connection with particle-decay from the giant dipole resonance (cf. Sect. 2.5). As already mentioned in connection with Fig. 2.8, the correlated 1p–1h giant resonance state excited in the inelastic scattering process, relaxes into more complex 2p–2h, 3p–3h, etc. states, eventually dissolving into the compound nucleus states. A simple approximation to this multistep picture is the two stage model used in Sect. 2.5, where the interacting states are the giant resonance and the compound states. Within this model, the cross section for the emission of γ-rays populating the ground state following inelastic scattering can be expressed as

$$\sigma_{x,x',\gamma_o}(E) = \sigma_{x,x'}(E)\left(\frac{\Gamma_{\gamma 0}}{\Gamma} + (\frac{\Gamma^{\downarrow}}{\Gamma})B_{CN}(E)\right). \qquad (2.14)$$

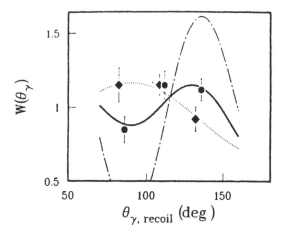

Figure 2.11. Experimental angular correlation of the ground state γ-decay from the excitation energy 9–11 MeV (circles) and 12–15 MeV (diamonds) in ^{208}Pb. The nucleus was excited with ^{17}O at 22 MeV/u and the angle for the detection of the high energy γ-ray was varied relative to the direction of the recoiling nucleus. The dashed curve is the prediction obtained in the framework of distorted wave Born approximation for a pure E1 decay from a 1^- state. The dashed-dotted curve corresponds to a pure E2 decay from a 2^+ state. The continuous curve represents the E2+E1 mixing predicted for the 9–11 MeV region. (Adapted from Beene et al. (1989).)

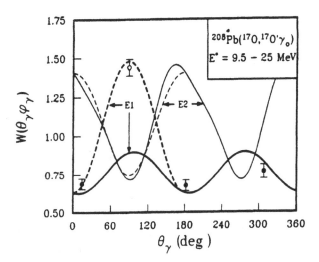

Figure 2.12. Experimental angular correlation of the γ-decay to the ground state of ^{208}Pb at excitation energy 9.5–25 MeV. The nucleus was excited with ^{17}O at 84 MeV/u and the angle for the detection of the high energy γ-ray was varied relative to the direction of the scattered ^{17}O nucleus measured at 2–3°. The lines are from theoretical calculations. Data points and continuous curves are associated to events lying in the reaction plane ($\phi = 0°$ and 180°). Open circle and dashed lines refer to the $\phi = 270°$ plane. Heavy curves refer to E1-deexcitation processes, while light curves refer to E2-transitions. (Adapted from Beene et al. (1990).)

In this expression, which displays a structure rather similar to that of Eq. (2.10), Γ is the experimental width (FWHM) of the resonance, while E is the nuclear excitation energy. The quantity $\sigma_{x,x'}(E)$ is the excitation cross section calculated within the framework of distorted wave Born approximation (DWBA). Making the ansatz that the giant resonance can be described as a sharp state at excitation energy E_{GR}, the value of the width Γ_{γ_0} can be written as

$$\Gamma_{\gamma_0} = g_I \frac{16\pi}{9(\hbar c)} E_{GR}^2 S_{GR}. \tag{2.15}$$

It contains the statistical factor $g_I = (2I_0 + 1)/(2I_R + 1)$ and the resonance oscillator strength S_{GR}. The quantities I_0 and I_R are the spin of the ground state and of the resonance state respectively. For a state exhausting 100% of the classical energy weighted sum rule strength (cf. Eq. (3.40)),

$$S_{GDR} = 14.8 \frac{NZ}{A} e^2 \text{fm}^2 \text{ MeV}. \tag{2.16}$$

The quantity B_{CN} appearing in Eq. (2.14) is the ground state γ-decay branch of the compound nucleus. It is calculated from mean ground state gamma-decay widths of the compound nucleus, and from widths calculated with the Hauser–Feshbach formalism (for details, see Hauser and Feshbach (1952)). The ratio $(\Gamma^\downarrow/\Gamma)$ ensures that only that fraction of the giant resonance which is damped among all the nuclear degrees of freedom is included in the compound term. Experimental and theoretical results for ^{208}Pb indicate that $\Gamma^\downarrow$ is $\approx 0.9\Gamma$ (cf. Chs. 4 and 10 and Sect. 2.5).

Figure 2.13(a) shows the experimental ground-state photon decay cross section associated with ^{208}Pb following inelastic scattering of ^{17}O at bombarding energy of 22 MeV/u. The data are compared with the predictions of the two-step model discussed above, assuming only the isovector giant dipole resonance to be present. The contributions to the total yield arising from the E1-decay of the giant resonance and of the compound nucleus are shown separately as dashed and dotted curves, respectively. One can notice that the dashed curve reflects the distribution of excitation cross section while the contribution from the compound nucleus (dotted curve) is much smaller for excitation energies larger than 12 MeV. This is because the ground state gamma-decay branch of the compound nucleus decreases with increasing excitation energy as the number of possible levels that can be populated in the compound decay increases. The calculations provide an overall account of the experimental data, exception made in the region near 10 MeV. This can better be seen from Figure 2.13(b), where the difference of the data and of the calculations is displayed in histogram form. The need for contributions arising from the decay of states different from the GDR is apparent. In fact, an overall account of the experimental findings is obtained taking into account, within the two-step γ-decay model, the contributions of L = 2 states at 8.8, 9.4 MeV (light solid line) and at 10.6 MeV (dashed curve), and assuming that these summed contributions

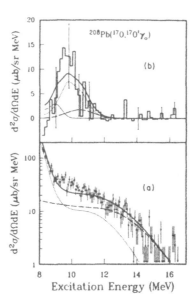

Figure 2.13. Ground-state gamma-decay coincidence data compared with calculations (heavy solid line) assuming only the presence of the giant dipole resonance and using the multistep theory of nuclear reactions (a). The GR contribution is shown with the dotted-dashed line while the compound contribution is shown with the short dashed line. The data are from the inelastic scattering of ^{17}O at 22 MeV/u. In part (b) the difference between the data and the calculated (heavy solid line) spectra displayed in part (a) is shown as histogram. The different spectrum shows a distinct peak located near the energy region of the isoscalar giant quadrupole resonance. The predicted (E2)γ_0 yield is shown in part (b) with the solid line. Also in this case the two contributions, GR and compound, are shown with the dotted-dashed and short dashed curves. (Adapted from Beene et al. (1989).)

exhaust 100 % of the isoscalar quadrupole energy weighted sum rule. The heavy solid curve is the total yield.

In contrast to dipole decay, the E2-ground state yield is dominated by the compound term (dotted curve). From the ratio of the inelastic cross section associated with the GQR (singles), and that associated with the ground-state gamma-decay coincidences, one obtains a GQR ground state branch of $4 \pm 1 \times 10^{-4}$.

The two-step decay formalism was also applied to the analysis of the ground state decay of the giant dipole resonance excited in the inelastic scattering of 84 MeV/u ^{17}O off ^{208}Pb. The results of the calculations are compared to the measurements in Fig. 2.14, where the separate contributions of the direct term (dotted curve) and of the compound term (dashed curve) are also shown. The energy-averaged total photon branch, integrated over the energy interval 9–25 MeV was found to be 0.017, as compared to the value of 0.019±0.002 deduced from the analysis of the angular correlation displayed in Fig. 2.12.

Recently, an effort was made to identify the two-phonon GDR states in ^{208}Pb and ^{209}Bi by γ-decay coincidence technique exploiting again the fact

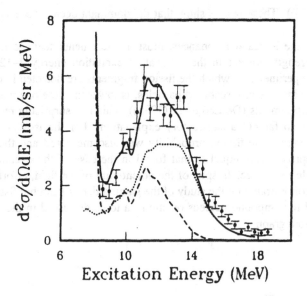

Figure 2.14. Ground-state gamma coincidence yield for 84 MeV/u ^{17}O scattering on ^{208}Pb compared with calculations. The solid curve is the result of the two-step model discussed in the text where both the direct and compound γ-decay are considered. The separate contribution of the two terms, direct (dotted curve) and compound (dashed line) are also shown. (Adapted from Beene et al. (1990).)

that the ground state γ-decay is dominated by E1-transitions (Beene (1994) and Ritman at al. (1993)). Photons from the decay of the two-phonon giant dipole resonance with energy $E_{2\hbar\omega_D} = 27$ MeV were identified. The yield of such photons was found to be a factor of ≈ 2 larger than that expected from theoretical models of the heavy ion excitation process, combined with simple assumptions concerning the γ-decay branching ratio of the two-phonon giant dipole resonance (cf., e.g., Ponomarev (1994)). Such discrepancy remains an open question. For a recent review of the situation concerning double-resonance physics cf. Chomaz and Frascaria (1995) and Emling (1994).

2.7 Fission Decay of Giant Resonances

The measurement of the fission width of giant resonances, in particular in the actinide region, provides information concerning the question of whether the decay takes place after or before the giant resonance is completely damped, and consequently following or not statistical laws.

In the case of giant dipole resonance at zero temperature, fission fragment measurements following γ-nuclear photoabsorption were made (Veyssiere et al. (1973), Caldwell et al. (1980)). The measured fission probability of $P_F(GDR) \approx 0.22$ was very similar to the values measured for compound nuclear reaction (n,f) and to what one expects from Z^2/A systematics (Bjørnholm and Lynn (1980)). These results show that the dominant decay mode is statistical.

Concerning the isoscalar resonances, most measurements seem to indicate that the E2-strength present in the interval of excitation energy 7–12 MeV measured in experiments in which the fission fragments are detected in coincidence with the inelastically scattered particles, is lower than the one measured in inclusive experiments (De Leo et al. (1985)). This is a surprising result for which there is so far not a satisfactory explanation. For the isoscalar giant monopole resonance the fission probability was also measured and the value found was approximately equal to that found in the case of the excitation of the giant dipole resonance. In spite of the potentiality of the field, efforts have been mainly concentrated on the study of the gamma-decay from hot fissioning nuclei formed in compound nucleus reactions, a topic discussed in the second part of this monograph.

3
Random Phase Approximation

The special stability displayed by certain nuclei, the so called closed shell nuclei, can be understood in terms of the filling of particular orbitals of the mean field where the nucleons move essentially independently of each other. On the other hand, nuclei can also react collectively. In a sense, all properties of a many-body system are collective. In particular, the existence of an average mean field is the most collective phenomenon displayed by the nucleus. While mean field properties are, as a rule, considered as opposed to collective properties, we shall see that both features have a common origin and emerge from the same components of the residual interaction (cf., e.g., Sects. 3.2.1 and 3.2.4), cf. also Mottelson (1962)). Furthermore, one can view the mean field as the organizing element of the variety of responses displayed by the nucleus to external fields. In fact, a through understanding of single-particle motion provides the basis for a microscopic description of vibrations and rotations.

3.1 Mean Field Theory

Mean field theory is a very useful approximation in the study of the many-particle system, in particular of the atomic nucleus. In it one replaces the many-particle Schrödinger equation

$$\left(-\sum_{i=1}^{A} \frac{\hbar^2}{2m} \nabla_i^2 + \sum_{i<j=1}^{A} v(|\vec{r}_i - \vec{r}_j|) \right) \Psi_n(\vec{r}_1 \ldots \vec{r}_A) = E_n \Psi_n(\vec{r}_1 \ldots \vec{r}_A),$$

(3.1)

sum of a kinetic energy term and a two-body interaction, by a single-particle Schrödinger equation:

$$\left(-\frac{\hbar^2}{2m} \nabla^2 + U(r) \right) \varphi_{\nu_i}(\vec{r}) + \int d^3 \vec{r}' U_x(\vec{r}, \vec{r}') \varphi_{\nu_i}(\vec{r}') = \varepsilon_{\nu_i} \varphi_{\nu_i}(\vec{r}). \quad (3.2)$$

The total wave function $\Psi_n(\vec{r}_1 \ldots \vec{r}_A)$ is here replaced by the normalized determinant constructed out of the single-particle wave functions $\varphi_i(\vec{r})$.

The two potentials appearing in Eq. (3.2) are the Hartree (direct) potential

$$U(\vec{r}) = \int d^3\vec{r'}\varrho(\vec{r'})v(|\vec{r} - \vec{r'}|), \qquad (3.3)$$

where

$$\varrho(\vec{r}) = \sum_{i=1}^{A} |\varphi_{\nu_i}(\vec{r})|^2,$$

is the total density of the system, and the Fock (exchange) potential

$$U_x(\vec{r}, \vec{r'}) = -\sum_{\nu_j=1}^{A} \varphi_{\nu_j}^*(\vec{r'})v(|\vec{r} - \vec{r'}|)\varphi_{\nu_j}(\vec{r}). \qquad (3.4)$$

This last term is directly connected with the fact that nucleons are fermions and thus satisfy the Pauli principle. In particular, the exchange potential takes care that nucleons do not interact with themselves.

The total energy of the system in the Hartree–Fock ground state $|0\rangle_{HF} = \frac{1}{\sqrt{A!}}\det(\varphi_{\nu_1}(\vec{r}_1)\ldots\varphi_{\nu_A}(\vec{r}_A))$ is given by

$$
\begin{aligned}
E = {}_{HF}\langle 0|H|0\rangle_{HF} &= \sum_{\nu_i \leq \nu_F} \langle \nu_i|T|\nu_i\rangle + \frac{1}{2}\sum_{\nu_i,\nu_{i'} \leq \nu_F} \langle \nu_i\nu_{i'}|v|\nu_i\nu_{i'}\rangle_a \\
&= \sum_{\nu_i \leq \nu_F} \varepsilon_i - \frac{1}{2}\sum_{\nu_i,\nu_{i'} \leq \nu_F} \langle \nu_i\nu_{i'}|v|\nu_i\nu_{i'}\rangle_a, \qquad (3.5)
\end{aligned}
$$

where ν_F labels the Fermi level lying, by definition (zero temperature situation), halfway between the last occupied and the first unoccupied orbitals. In writing up the last term of the above equation the self-consistency relation

$$
\begin{aligned}
\langle \nu_2|T|\nu_1\rangle &+ \sum_{\nu_i \leq \nu_F} \langle \nu_i\nu_2|v|\nu_i\nu_1\rangle_a \\
&= \langle \nu_2|T + U + U_x|\nu_1\rangle = \varepsilon_{\nu_1}\delta(\nu_1,\nu_2),
\end{aligned}
$$

that is, the matrix expression of Eq. (3.2), has been used. Note that the subindex a in the matrix element indicates the antisymmetrized matrix element, i.e., $\langle \nu_i\nu_k|v|\nu_i\nu_k\rangle_a = \langle \nu_i\nu_k|v|\nu_i\nu_k\rangle - \langle \nu_i\nu_k|v|\nu_k\nu_i\rangle$, and thus gives rise to both the direct and exchange potentials. The factor $\frac{1}{2}$ in the last term of Eq. (3.5) reflects the fact that the two-particle interaction contributes to the average potential for both of the interacting particles and is thus counted twice, if one sums the single-particle energies for the filled orbitals.

3.1.1 Effective mass (k−mass)

Experimental evidence testifies to the fact that single-particle motion in nuclei is well described by a potential of Woods–Saxon type:

$$U(r, E) = V_0(E) f(r), \tag{3.6}$$

where

$$f(r) = \frac{1}{1 + \exp(\frac{r - R_0}{a})}, \tag{3.7}$$

to which a spin-orbit potential, proportional to $\frac{\partial f(r)}{\partial r}$, is added (cf. Bohr and Mottelson (1969)). The radius and the diffuseness parameters have the values

$$R = r_0 A^{\frac{1}{3}} \text{ fm}, \quad r_0 = 1.25 \text{ fm}, \quad a = 0.65 \text{ fm}, \tag{3.8}$$

and for levels around the Fermi energy the strength $V_0(E)$ is a constant. On the other hand, the differential elastic scattering cross section and the total nucleon–nucleus cross section can be accurately described with

$$V = V_0(E) = V^0 + \gamma E, \tag{3.9}$$

with $V^0 = -55$ MeV and $\gamma = 0.3$–0.4, provided that one adds to the potential given in Eq. (3.6) an imaginary component. We shall come back to this question in Sect. 4.1. The same parametrization describes the deeply bound states as shown in Fig. 3.1.

The relation (3.9) is valid for $|E| > 10$ MeV, where the single-particle energy E $(= \varepsilon - \varepsilon_F)$ is measured from the Fermi energy. The valence orbitals $(|E| \leq 5$ MeV) of nuclei around closed shells are well reproduced by the Woods–Saxon potential defined by Eqs. (3.6)–(3.8) but in this case with $V \approx -55$ MeV, independent upon energy.

The Schrödinger equation

$$\left(-\frac{\hbar^2}{2m} \nabla^2 + U(r, \varepsilon_j) \right) \varphi_j(\vec{r}) = \varepsilon_j \varphi_j(\vec{r}). \tag{3.10}$$

can, for many purposes, be rewritten, to a very good approximation, as

$$\left(-\frac{\hbar^2}{2m^*} \nabla^2 + \tilde{U}(r) \right) \tilde{\varphi}_j(\vec{r}) = \tilde{\varepsilon}_j \tilde{\varphi}_j(\vec{r}), \tag{3.11}$$

where the concept of effective mass

$$m^* = m \left(1 - \frac{dV_0}{d\varepsilon} \right) \tag{3.12}$$

has been introduced. The depth of the energy-independent potential $\tilde{U}(r)$ is $\simeq mV^0/m^*$, the relation being exact for an infinite uniform system, without radial dependence. In Hartree–Fock theory, contributing to the energy dependence of the single-particle potential are the non-locality of $U_x(\vec{r}, \vec{r}')$, equivalent to a dependence on the linear momentum of the particle and, in many

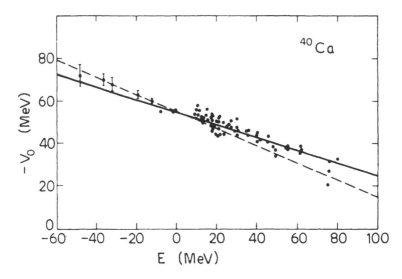

Figure 3.1. Dependence upon proton bombarding energy of the depth V of the potential well (3.6) which reproduces the ^{40}Ca(p,p) differential cross section for $E > 10$ MeV, and the experimental single-proton energies for $E < 0$ MeV. The full straight line corresponds (in MeV) to $V = V_0(E) = -55$ MeV $+ 0.3E$, and the dashed straight line to $V = V_0(E) = -55$ MeV $+ 0.4E$. (Adapted from M. Bauer et al. (1982).)

cases, the genuine velocity dependence of the two-body interaction. Equation (3.12) with the parametrization given in Eq. (3.9) leads to an effective mass, known as the k-mass, which is considerably smaller than the bare nuclear mass[1]. In fact,

$$m_k \approx 0.6m - 0.7m(= m^*). \tag{3.13}$$

Consequently, Hartree–Fock theory is able to accurately predict the sequence of the single-particle levels around the Fermi energy (i.e., $|E| \leq 10$ MeV), but not its density. This is exemplified in Fig. 3.2, where the experimental values of the single-particle neutron energies of the valence orbitals of ^{208}Pb, are compared with Hartree–Fock results calculated making use of a particular parametrization of the effective two-body interaction v known as a Skyrme interaction.

The limitation mentioned above of Hartree–Fock theory is connected with the fact that one is dealing with a static approximation to the many-body problem. In fact, the presence of a mean field defines a surface. This surface can vibrate. These vibrations renormalize the properties of the mean field leading to an increase in the nuclear radius and in the diffusivity. Consequently,

---

[1] In an infinite system, where the Hartree–Fock energy-momentum dispersion relation reads $E = \hbar^2 k^2/2m + V_0^{HF}(k)$, one obtains $m^*/m \equiv m_k/m, = (1 + \frac{m}{\hbar^2 k} dV_0/dk)^{-1}$.

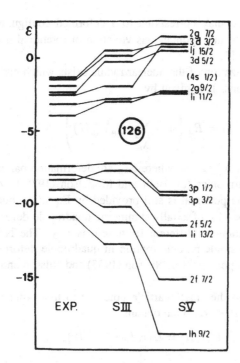

Figure 3.2. Comparison between the experimental single-neutron energies in the valence shells of ^{208}Pb and the values calculated in the Hartree–Fock approximation with a Skyrme III (SIII, middle) and a Skyrme V (SV, right) interaction. (Taken from Quentin and Flocard (1978).)

levels with an energy somewhat larger than the Fermi energy will decrease their energy coming closer in energy to ε_F. Levels below the Fermi energy will feel two contrasting effects. The increase in the radius will decrease their kinetic energy while the increase in the diffusivity will increase it. In any case, the net result is an increase of the density of levels around the Fermi energy (cf. Sect. 4.1.2) which can be viewed in terms of an increase of the effective mass m_k to the bare mass m, as required by the experimental findings. The energy parameter controlling these effects is the energy of the low-lying surface vibrations. This quantity is, in general, of the order of 1–3 MeV. Consequently, single-particle levels with energy $|E|$ much larger than $\approx$ 5 MeV will not be affected by surface vibrations, and will display the effective mass given in Eq. (3.13).

3.2 Random Phase Approximation (RPA)

In solving the Hartree–Fock equations one has to specify the shape of the nucleus. For closed shell systems, the absolute minimum of the energy of the system is associated with a spherical configuration. For nuclei with a number of nucleons outside closed shell, or a number of holes in the closed shell, the absolute minimum may correspond to a deformed configuration. Some of the consequences of this spontaneous violation of rotational invariance are discussed in Chapter 11.

A simple parametrization of the nuclear radius exists which can account for the variety of situations. It is given by

$$R = R_0 \left(1 + \sum_{\lambda\mu} \alpha_{\lambda\mu} Y_{\lambda\mu}^*(\hat{r}) \right), \tag{3.14}$$

with the multipolarity $\lambda \geq 2$ and where $\alpha_{\lambda\mu}$ are deformation parameters while $Y_{\lambda\mu}$ are spherical harmonics (cf. Bohr and Mottelson (1975)). An adequate parametrization of the potential is still provided by Eq. (3.7), but with R_0 is replaced by R. In the case of axially symmetric quadrupole deformations, the only deformation parameter different from zero is α_{20}. The Nilsson model used to describe the single-particle motion in quadrupole deformed nuclei is closely related to this potential (cf. Nilsson (1955) and Nilsson and Ragnarsson (1995)).

Let us now expand the single-particle potential to first order in the deformation parameters ($\alpha^2 \ll \alpha$). One obtains

$$U(r, R) = U(r, R_0) + \delta U(r), \tag{3.15}$$

where

$$\delta U(r) = -R_0 \frac{\partial U}{\partial r} \sum_{\lambda\mu} \alpha_{\lambda\mu} Y_{\lambda\mu}^*(\hat{r}). \tag{3.16}$$

It is well established that the nuclear surface can vibrate in certain normal modes. In this case the quantities $\alpha_{\lambda\mu}$ can be viewed as the coordinates of the harmonic oscillator Hamiltonian associated with the normal modes, that is,

$$H_\alpha = \frac{\hat{\Pi}_\alpha^2}{2D_\alpha} + \frac{C_\alpha}{2}\hat{\alpha}^2, \tag{3.17}$$

where

$$\hat{\alpha} = \sqrt{\frac{\hbar\omega_\alpha}{2C_\alpha}}(\hat{\Gamma}_\alpha^\dagger + \Gamma_\alpha), \tag{3.18}$$

and $\hat{\Pi}_\alpha$ is the variable conjugated to $\hat{\alpha}$. The quantities $\Gamma_\alpha^\dagger$ and Γ_α are boson creation and annihilation operators (Dirac (1935)) of the vibrational modes α, that is,

$$|\alpha\rangle = \Gamma_\alpha^\dagger |0\rangle_B, \tag{3.19}$$

$|0\rangle_B$ being the boson vacuum. The quantity $(\frac{\hbar\omega_\alpha}{2C_\alpha})^{1/2}$ is the zero point fluctuation associated with the mode. Consequently, the term δU leads to a coupling between the single-particle motion described in terms of the coordinate $\vec{r}$, and the collective vibrations, described in terms of the collective coordinates $\hat{\alpha}$, which we write as

$$\delta U = -\kappa \hat{\alpha} \hat{F}. \tag{3.20}$$

The dimensionless quantity

$$\hat{F} = \frac{R_0}{\kappa} \frac{\partial U}{\partial r} Y^*_{\lambda\mu}(\hat{r}), \tag{3.21}$$

is a single-particle field peaked at the nuclear surface. In a normal self-sustained mode, there should be a consistency between variations of the density and of the potential. As we shall see (cf. Sects. 3.2.3 and 3.2.5), the quantity κ is the proportionality constant between these two variations.

We are here treating angular momentum in a very superficial way. This is done on purpose to be able to discuss the main physical consequences of the particle-vibration coupling Hamiltonian defined in Eq. (3.20) in simple terms. We refer to Bohr and Mottelson (1975) and to Bortignon et al. (1977) for the detailed expressions containing the proper angular momentum coupling coefficients.

The basic process described by the particle-vibration coupling Hamiltonian δU is that of a particle scattering inelastically off the surface and setting it into vibration, as shown in Fig. 3.3. The ease with which the process takes place is measured by the matrix element

$$V(\nu_k, \nu_{k'}; \alpha) = \langle \alpha \nu_{k'} | \delta U | \nu_k \rangle = \Lambda_\alpha \langle \nu'_k | \hat{F} | \nu_k \rangle, \tag{3.22}$$

where

$$\Lambda_\alpha = -\kappa \sqrt{\frac{\hbar \omega_\alpha}{2 C_\alpha}} = \frac{-\kappa \beta_{\lambda\alpha}}{\sqrt{2\lambda_\alpha + 1}}, \tag{3.23}$$

is the strength with which the particle couples to the vibration, while

$$\langle \nu'_k | \hat{F} | \nu_k \rangle = \int d^3r \, \varphi^*_{\nu_{k'}}(\vec{r}) F(\vec{r}) \varphi_{\nu_k}(\vec{r}), \tag{3.24}$$

$\varphi_{\nu_k}(\vec{r})$ and $\varphi_{\nu'_k}(\vec{r})$ being the single-particle wave functions, solutions of Eqs. (3.11) and (3.13).

The quantity $\beta_{\lambda\alpha}$ is associated with the deformation parameters introduced in Eq. (3.14). In particular for $\lambda = 2$ and $\mu = 0$, we have $\alpha_{20} = \beta_2$ (cf. also Eq. (2.2)). A similar matrix element can be obtained when the fermion, instead of a particle above the Fermi surface, is a hole in the Fermi sea (we refer to Sect. 4.2 (cf. Eqs. (4.34) and (4.35)) for a discussion of the relation between the corresponding matrix element and the matrix element (3.22)). Aside from these matrix elements, the particle-vibration coupling Hamiltonian allows for the following two matrix elements (cf. Fig. 3.4):

$$\langle \alpha | \delta U | \nu_k (\nu_i)^{-1} \rangle = \Lambda_\alpha \langle \tilde{\nu}_i | \hat{F} | \nu_k \rangle, \tag{3.25}$$

and

$$\langle \alpha \nu_k (\nu_i)^{-1} | \delta U | 0 \rangle = \Lambda_\alpha \langle \tilde{\nu}_i | \hat{F} | \nu_k \rangle^*, \tag{3.26}$$

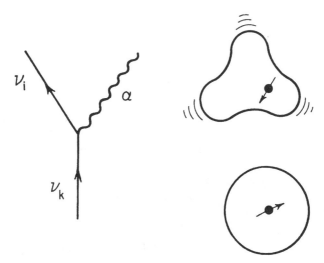

Figure 3.3. Graphical representation of the process by which a fermion, bouncing inelastically off the surface sets it into vibration. Particles are represented by an arrowed line, while the vibration is shown by a wavy line. The black dot represents a nucleon moving in a spherical mean field of which it excites an octupole vibration after bouncing inelastically off the surface.

where the symbol $|\nu_i^{-1}\rangle$ denotes a hole state while $|\tilde{\nu}_i\rangle$ is the state time-reversed to the state $|\nu_i\rangle$. The first matrix element describes the process by which a particle falls into a hole giving its energy and angular momentum to a vibrational state $|\alpha\rangle$. The second matrix element displayed is associated with the process by which the vacuum becomes virtually excited through the simultaneous presence of a particle, of a hole and of a vibration.

3.2.1 Dispersion relation

Through the particle-vibration coupling Hamiltonian δU it is possible to calculate the normal modes of a system. This can be done by recognizing the dual character of Eq. (3.20), in the sense that a collective mode can be excited through the field $\hat{\alpha}$ as well as through the field $\hat{F}$. In other words, a consequence of the coupling δU is that a collective vibration can be viewed as a correlated particle–hole excitation, which, in the independent particle basis corresponds to a linear combination of particle–hole excitations. Consequently, in the particle-vibration coupling model, the transition amplitude

$$\langle \alpha | \hat{\alpha} | 0 \rangle = \sqrt{\frac{\hbar \omega_\alpha}{2 C_\alpha}}, \tag{3.27}$$

should be equal to (cf. Fig. 3.5)

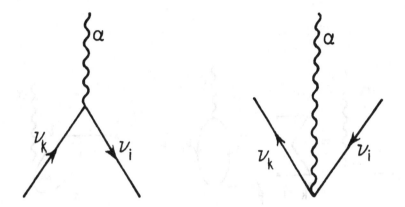

Figure 3.4. Graphical representation of two matrix elements of δU.

$$\langle \alpha | \hat{F} | 0 \rangle$$

$$= \sum_{\nu_k, \nu_i} \left\{ \frac{\langle \alpha | \delta U | \nu_k \nu_i^{-1} \rangle \langle \nu_k \nu_i^{-1} | \hat{F} | 0 \rangle}{\hbar \omega_\alpha - (\varepsilon_{\nu_k} - \varepsilon_{\nu_i})} \right.$$

$$\left. + \frac{\langle \alpha | \hat{F} | \nu_k \nu_i^{-1}; \alpha \rangle \langle \nu_k \nu_i^{-1}; \alpha | \delta U | 0 \rangle}{-(\hbar \omega_\alpha + (\varepsilon_{\nu_k} - \varepsilon_{\nu_i}))} \right\}$$

$$= - \sum_{\nu_k, \nu_i} (X_\alpha(\nu_k \nu_i) + Y_\alpha(\nu_k \nu_i)) \langle \tilde{\nu}_i | \hat{F} | \nu_k \rangle, \qquad (3.28)$$

where

$$\left. \begin{array}{c} X_\alpha(\nu_k \nu_i) \\ Y_\alpha(\nu_k \nu_i) \end{array} \right\} = \pm \frac{\Lambda_\alpha \langle \tilde{\nu}_i | \hat{F} | \nu_k \rangle}{(\varepsilon_{\nu_k} - \varepsilon_{\nu_i}) \mp \hbar \omega_\alpha}. \qquad (3.29)$$

For simplicity, the matrix element $\langle \tilde{\nu}_i | \hat{F} | \nu_k \rangle$ has been assumed to be real. Equating the relations given in Eqs. (3.27) and (3.28) one obtains the dispersion relation

$$W(\hbar \omega_\alpha) = \sum_{\nu_k, \nu_i} \frac{2(\varepsilon_{\nu_k} - \varepsilon_{\nu_i}) |\langle \tilde{\nu}_i | \hat{F} | \nu_k \rangle|^2}{(\varepsilon_{\nu_k} - \varepsilon_{\nu_i})^2 - (\hbar \omega_\alpha)^2} = \frac{1}{\kappa}. \qquad (3.30)$$

This relation can be solved numerically for the values of $\hbar \omega_\alpha$ as illustrated in Fig. 3.6.

In keeping with the relation given in Eq. (3.28) one can write the phonon creation operator as

$$\Gamma_\alpha^\dagger = \sum_{\nu_k, \nu_i} X_\alpha(\nu_k, \nu_i) \Gamma_{\nu_k \nu_i}^\dagger + Y_\alpha(\nu_k, \nu_i) \Gamma_{\nu_k \nu_i}, \qquad (3.31)$$

where $\Gamma_{\nu_k \nu_i}^\dagger = a_{\nu_k}^\dagger a_{\nu_i}$ and $\Gamma_{\nu_k \nu_i} = (a_{\nu_k}^\dagger a_{\nu_i})^\dagger$ are creation and annihilation operators of pairs of fermions which are assumed to display boson commutation

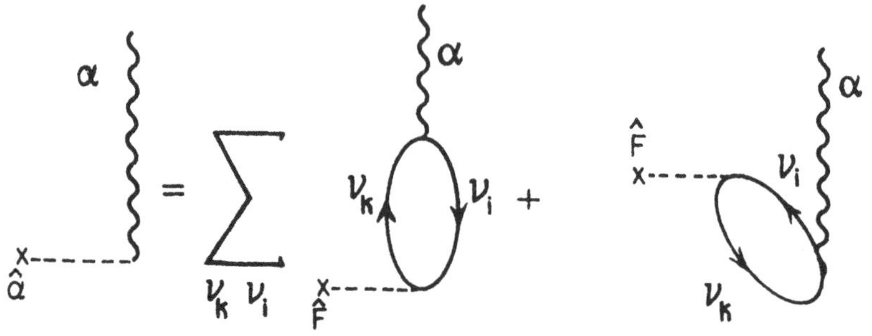

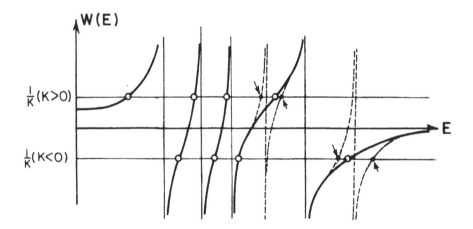

Figure 3.5. Excitation of the collective vibration in terms of the operators $\hat{\alpha}$ and $\hat{F}$. (After Bohr and Mottelson (1975).)

Figure 3.6. Graphical solution of the RPA dispersion relation, Eq. (3.30). The dashed lines and curves exemplify the phenomenon of the decay of giant vibrations into particle motion (Landau damping), while the arrows point to the corresponding roots.

relations as $\Gamma_\alpha^\dagger$ and Γ_α do. This is the essence of the so called Random Phase Approximation (RPA). Consequently,

$$1 = [\Gamma_\alpha, \Gamma_\alpha^\dagger] = \sum_{\nu_k, \nu_i} (X_\alpha^2(\nu_k, \nu_i) - Y_\alpha^2(\nu_k, \nu_i)), \qquad (3.32)$$

a relation which ensures that the one-phonon state $|\alpha\rangle = \Gamma_\alpha^\dagger |0\rangle_B$ is normalized. Eq. (3.32) provides the following microscopic expression for the square of the particle-vibration coupling strength

$$\Lambda_\alpha^2 = \left\{ 2\hbar\omega_\alpha \sum_{\nu_k, \nu_i} \frac{2(\varepsilon_{\nu_k} - \varepsilon_{\nu_i})|\langle \tilde{\nu}_i | \hat{F} | \nu_k \rangle|^2}{[(\varepsilon_{\nu_k} - \varepsilon_{\nu_i})^2 - (\hbar\omega_\alpha)^2]^2} \right\}^{-1}$$

$$= \left(\frac{\partial W(E)}{\partial E} \Big|_{E=\hbar\omega_\alpha} \right)^{-1}. \qquad (3.33)$$

Because $(\Lambda_\alpha/\kappa)^2 = (\hbar\omega_\alpha/2C_\alpha)$, the above relation also provides the value of the transition probability $\langle \alpha | \hat{F} | 0 \rangle^2$.

Let us comment on the general features of the graphical solutions of Eq. (3.30), as schematically displayed in Fig. 3.6. The poles of the dispersion relation $W(E)$ correspond to the values of the particle–hole excitation energies. Each root $\hbar\omega_\alpha$ is, in general, bound by two poles and there are as many states $|\alpha\rangle$ as particle–hole states $|\nu_k \nu_i^{-1}\rangle$. The collectivity of a state $|\alpha\rangle$ is measured by the inverse of the derivative of the dispersion relation $W(E)$ with respect to E at the value $E = \hbar\omega_\alpha$ (cf. Eq. (3.33)). Consequently, roots which are bound by two poles with similar energies, will display little collectivity, as the associated derivative at the corresponding root is very large. Because of this, a single amplitude $X_\alpha(\nu_k, \nu_i)$ will dominate the microscopic structure of the associated wavefunction (cf. Eqs. (3.19), (3.29) and (3.31)). Collective modes are possible when there is a gap in the particle–hole excitation spectrum. This can happen either at low excitation energies ($\leq$ 3–4 MeV), and is in general connected in medium-heavy nuclei with the spin-orbit splitting of single-particle levels, or at high excitation energies in connection with the energy separation between major shells.

We are now in a position to clarify the meaning of the statement made at the beginning of this chapter, concerning the fact that the mean field is the organizing element in the study, not only of single-particle motion, but also of vibrational motion. In fact, the graphical solutions of Eq. (3.30) show that the properties of the vibrational modes are directly controlled by the detailed distribution of single-particle levels. For $\kappa < 0$ (repulsive particle-vibration coupling strength), there are no low-lying collective vibrations and the only collective modes (that is, small $\frac{\partial W(E)}{\partial E}|_{\hbar\omega_\alpha}$) are found at high excitation energies, associated with particle–hole excitations across major shells, displaying tens of MeV of energy. These modes can be excited by fields which change very fast in time ($\hbar/(10 \text{ MeV}) \sim 10^{-22}$ sec). They represent the elastic response of the system to instantaneous solicitations (cf. Sect. 1.4). For $\kappa \geq 0$ (attractive particle-vibration coupling strength), added to these high lying modes, vibrations with low excitation energy (1–2 MeV) are possible. These vibrations are a manifestation of the plastic nuclear response (cf. Sect. 1.5) to external fields which change slowly in time ($\hbar/(1$ MeV$) \sim 10^{-21}$ sec). For the value of κ for which the lowest root of the dispersion relation becomes zero, the spherical configuration is no longer stable and the system becomes deformed. Situations of this type happen, as a rule, for nuclei far away from closed shell and for vibrations of spin and parity 2^+. Consequently, the above situation corresponds to a quadrupole vibration degenerate with the ground state. This result not only implies that one is applying the RPA model beyond its region of validity, but that a new phenomenon is taking place.

The assumption which is at the basis of the RPA is small amplitude motion (cf. Eq. (3.16), $\alpha^2 \ll \alpha$) and consequently harmonic vibrations. Because $\hbar\omega_\alpha = (C_\alpha/D_\alpha)^{\frac{1}{2}}$, if a root becomes zero, $C_\alpha = 0$. This implies that the restoring force of the system with respect to deformations, in particular quadrupole deformations, vanishes and the nucleus becomes permanently deformed. This "phase transition", as exemplified in Fig. 1.5 by the spectrum of Sm-isotopes as a function of nucleon number, implies a violation of the harmonic assumption and the inability of the RPA to deal with such a situation. Consequently, the absolute minimum in the ground state energy corresponds to a Hartree–Fock solution where the system displays a static deformation. The fact that a solution of the spherically symmetric Hamiltonian is not rotational invariant is an example of the phenomenon of spontaneous symmetry breaking. Harmonic vibrations can still take place around this new stable shape (β- and γ-vibrations, that is the states at 1110 keV and 1444 keV in ^{154}Sm respectively in Fig. 1.5). The restoration of rotational symmetry is obtained by fluctuations of the orientation of the system, leading to quadrupole rotational bands, as displayed by ^{154}Sm.

3.2.2 Sum rules

The Random Phase Approximation provides a diagonalization of the particle-vibration coupling Hamiltonian within the harmonic approximation. It is then natural that, as stated before, the number of states $|\alpha\rangle$ is equal to the number of particle–hole states $|\nu_k \nu_i^{-1}\rangle$ coupled to the quantum numbers of the vibration and which form the basis states. Provided that the interaction acting among the fermions is velocity independent, the product of the energy of these states times the square of matrix elements between a particle and a hole state of any one-body operator which only depends on the spatial coordinate is a model independent quantity, reflecting very simple properties of the system as a whole. This result is known as an energy weighted sum rule (EWSR). In the case of dipole excitations it is proportional to the total number of charged particles of the system, being also proportional to the photoabsorption cross section. One of the basic conditions to be fulfilled by any theoretical treatment used to diagonalize the residual interaction acting between particle–hole states should be to conserve this sum rule.

The basic importance of sum rules in the study of vibrational motion is that they are connected with basic operator identities which restrict the possible matrix elements in a physical system. Also, through the use of sum rules it is possible to assess the collectivity of a given excitation. Finally, sum rules provide an upper limit to the energy that can be transferred to a nucleus under the action of an external field (Broglia and Winther (1991)). The subject of sum rules is quite general and in what follows we shall only touch upon it. In particular we shall discuss sum rules associated with spatial dependent single-particle fields.

It is simple to show that

$$\sum_n |\langle 0|\hat{F}|n\rangle|^2 (E_n - E_0) = \frac{1}{2}\langle 0|[\hat{F}, [H, \hat{F}]]|0\rangle, \qquad (3.34)$$

where n labels the complete set of eigenstates of H, $|0\rangle$ being the ground state. If $\hat{F} = \Sigma_k F(\vec{r}_k)$ is a one-particle operator, depending only on the spatial coordinates, the double commutator receives only contributions from the kinetic energy part of the Hamiltonian, that is,

$$\begin{aligned}
\frac{1}{2}\langle 0|[\hat{F}, [H, \hat{F}]]|0\rangle &= \frac{1}{2}\langle 0|[\hat{F}, [T, \hat{F}]]|0\rangle \\
&= \langle 0| \sum_k \frac{\hbar^2}{2m}(\vec{\nabla}_k \hat{F}(\vec{r}_k))^2|0\rangle, \qquad (3.35)
\end{aligned}$$

where the last term implies the diagonal matrix element in the ground state. The commutativity of the operator $\hat{F}$ depending only on the spatial coordinates with the nuclear interaction is true for a large class of nuclear potentials, including zero range velocity dependent interactions of Skyrme type, as discussed below. The average in Eq. (3.35) can be replaced by an integral over the density, that is

$$\sum_n |\langle 0|\hat{F}|n\rangle|^2 (E_n - E_0) = \frac{\hbar^2}{2m} \int d^3r \, |\vec{\nabla}\hat{F}|^2 \varrho(\vec{r}). \qquad (3.36)$$

This simple result is what one would expect from the reaction of a system in equilibrium to which one applies an impulsive field, which gives the particles a momentum $\vec{\nabla}\hat{F}$. On the average, the particles started at rest so their average energy after the sudden impulse is $\hbar^2|\vec{\nabla}\hat{F}|^2/2m$. This result is consistent with the fact that the energy weighted sum rule does not depend on the interactions acting among the nucleons, as the energy is absorbed before the system is disturbed from equilibrium.

The presence of a velocity dependence (of an effective mass, cf. Eq. (3.13) and preceding discussion) in the mean field and thus a violation of Galilean invariance, requires some further discussions concerning the value of the mass to be used in the above equations. This violation must be corrected for by adding to δU an additional particle-vibration coupling field (Bohr and Mottelson (1975))

$$\delta U_1 \propto \dot\alpha \frac{\partial V_0(E)}{\partial p_z}, \tag{3.37}$$

resulting from a uniform collective motion with velocity $\dot\alpha$ in, for example, the z-direction, relative to which the velocity of the nucleons which appears in $V_0(E)$ defined in Eq. (3.9) should be measured. The proportionality coefficient for a given nuclear interaction is such as to cancel, for an isoscalar field F, the extra contribution to the commutator produced by the velocity dependence of the mean field. Consequently, in this case the bare nucleon mass appears in Eqs. (3.35) and (3.36). This is not the case for an isovector field F, essentially because the neutron–proton interaction is different from the interaction between identical nucleons. Therefore, the result of Eq. (3.35) is modified by a factor which is usually written as $(1 + K)$, where K depends on the interaction, on the multipolarity of F and on the nucleus (cf., e.g., Lipparini and Stringari (1989) and refs. therein). Because of the factor $(1 + K)$, the energy weighted sum rule has a different value at the level of RPA than of Hartree–Fock. In this chapter we shall disregard this factor and shall use the bare mass, as if the potential was velocity independent. We shall come back to this subject in Sect. 10.3.

The energy weighted sum rule most often used in the case of finite systems is that associated with multipole fields, $F(\vec{r}) = r^L Y_{LM}(\hat{r})$. In this case,

$$\begin{aligned} &\sum_n \ |\langle 0|r^L Y_{LM}|n\rangle|^2 (E_n - E_0) \\ &= \frac{\hbar^2}{2m} \frac{(2L+1)^2 L}{4\pi} \int d^3r \ r^{2L-2} \varrho. \end{aligned} \tag{3.38}$$

The most important application of the sum rule identities is to the case of a constant force field, that is in the case where F has a constant gradient. The electric field from the photon is of this form in the dipole approximation, which is valid when the size of the system is small compared to the wavelength of the photon. Inserting (cf. Sect. 10.1)

$$\hat{F}(\vec{r}_k) = e\Big[\frac{N-Z}{A} - t_z(k)\Big] r_k Y_1 \mu(\vec{r}_k), \tag{3.39}$$

with $t_z = -1/2$ for protons and $+1/2$ for neutrons, for the dipole operator referred to the nucleus centre-of-mass in Eq. (3.36), one obtains

$$\sum_n |\langle 0|\hat{F}|n\rangle|^2 (E_n - E_0) = \frac{.9}{4\pi} \frac{\hbar^2 e^2}{2m} \frac{NZ}{A}. \tag{3.40}$$

This is known as the Thomas–Reiche–Kuhn sum rule.

3.2.3 Coupling strength

In a self-sustained vibration the changes in the density should be proportional to the changes in the potential. The coupling constant κ in Eq. (3.20) provides this proportionality factor. As previously discussed, the operators $\hat{\alpha}$ and $\hat{F}$ can be viewed as the collective and the single-particle representation of the same field. In other words, Eq. (3.20) can also be thought in terms of a two-body residual interaction

$$v(\vec{r}, \vec{r}') = -\kappa_1 \hat{F}(\vec{r}) \hat{F}(\vec{r}'). \qquad (3.41)$$

Let us first discuss the case of a dipole field of the form $\tau_z z$, acting on both protons ($\tau_z = -1$) and neutrons ($\tau_z = +1$). If the isospin factor was not present, the field would act equally on protons and neutrons leading to a shift of the center of mass but not to an internal oscillation of the nucleus. Under the action of the field $F = \tau_z z$, protons and neutrons will oscillate with opposite phases (cf. Fig. 1.2), leading to an internal vibration of the system to which is associated a change in the isovector density

$$\delta\varrho_1 = cz\varrho, \qquad (3.42)$$

where ϱ is the total density of the system. This change in the isovector density will induce a change in the isovector potential given by

$$
\begin{aligned}
\delta\mathcal{U}_1 &= \int d^3r'\, \delta\varrho_1 v = -\kappa_1 cz \int d^3r'\, z'^2 \varrho \\
&= -c\kappa_1 zA\langle z^2\rangle = -c\kappa_1 z \frac{A}{3} \langle r^2\rangle,
\end{aligned} \qquad (3.43)
$$

where

$$\langle r^L\rangle = \frac{\int d^3r\, r^L \varrho_0}{\int d^3r\, \varrho_0} \approx \frac{3}{3+L} R_0^L. \qquad (3.44)$$

Equation (3.43) is the extension, to the dynamic case, of the self-consistent relation existing between the equilibrium density and the potential, as given by Eq. (3.3). The single-particle potential depends on isospin approximately as (cf., e.g., Bohr and Mottelson (1969))

$$U(r, E) = \left(V_0(E) + \tau_z \frac{N-Z}{A} V_1\right) f(r), \qquad (3.45)$$

where $V_0(E)$ has been defined in Eq. (3.9). To proceed further we need to calculate the value of the strength V_1 of the isospin potential. For this purpose we make use of the fact that total potential energy can be written in the independent particle approximation, as one-half the sum of the one-particle potential energy. Using for simplicity $f(r) \approx f(0) \approx 1$ (square-well potential), one obtains

$$V_{sym} \approx \frac{1}{2} \frac{(N-Z)^2}{A} V_1. \qquad (3.46)$$

This formula can be compared with the isospin term of the semi-empirical mass formula (Weiszäcker, (1935))

$$\mathcal{E}_{sym} = \frac{1}{2} b_{sym} \frac{(N - Z)^2}{A},$$

where $b_{sym} \approx 50$ MeV. The symmetry energy may be divided into a kinetic and potential energy part. A simple estimate of the kinetic energy part can be obtained by making use of the Fermi gas model which gives

$$(b_{sym})_{kin} \approx \frac{2}{3}\varepsilon_F \approx 25 \text{ MeV}. \tag{3.47}$$

Consequently,

$$V_1 = (b_{sym})_{pot} = b_{sym} - (b_{sym})_{kin} \approx 25 \text{ MeV}. \tag{3.48}$$

Writing the variation of the potential induced by a variation of the isovector density in the form (cf. Eq. (3.45))

$$\delta \mathcal{U}_1 = V_1 \frac{\delta \varrho_1}{\varrho} = V_1 cz, \tag{3.49}$$

and equating this relation to Eq. (3.43) one obtains

$$\kappa_1 = -\frac{3V_1}{A\langle r^2 \rangle}, \tag{3.50}$$

for the isovector dipole coupling constant. Making use of the relation given in Eq. (3.44) for $L = 2$ one can write

$$\kappa_1 = -\frac{5V_1}{AR_0^2}. \tag{3.51}$$

The negative sign implies a repulsive interaction, in keeping with the fact that it takes energy to separate protons from neutrons.

3.2.4 Frequency of dipole vibrations

We shall now solve the dispersion relation given in Eq. (3.30) for the case of isovector dipole vibrations, that is we shall assume $\hat{F} = z\tau_z$. Furthermore, we shall use an harmonic oscillator potential to describe the single-particle motion. This potential provides a simple and in many cases adequate description of the properties of single-particle orbitals. In the present estimates we shall furthermore neglect any spin-orbit interaction. The neglect of spin-orbit terms is of no consequences for giant vibrations. However, this approximation is not valid in the case one is interested in low-lying modes, i.e., vibrations at energies considerably smaller than $\hbar\omega_0$ ($\approx 41/A^{\frac{1}{3}}$), where ω_0 is the frequency of the oscillator adjusted to fit the Saxon–Woods potential (cf. Fig. 3.7) defined in Eqs. (3.6)–(3.9) setting $\gamma = 0$.

Because the matrix elements of the coordinate z are different from zero only when the states connected differ by one in the value of the principal

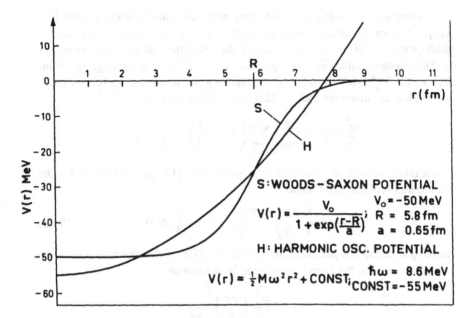

Figure 3.7. Comparison of the Saxon–Woods potential defined in Sect. 3.1 and of an harmonic oscillator potential whose frequency has been chosen in order to fit the Saxon–Woods potential. (From Bohr and Mottelson (1969).)

quantum number, there is only one particle–hole excitation energy, namely, $\varepsilon_{\nu_k} - \varepsilon_{\nu_i} = \hbar\omega_0$. Within this approximation, the sum in Eq. (3.30) affects only the numerator, which can be replaced by the sum rule given in Eq. (3.36). For the dipole operator $F = z\tau_z$, this quantity is equal to $(\hbar^2 A/2m)$. Thus

$$(\hbar\omega_D)^2 = (\hbar\omega_0)^2 + \frac{3\hbar^2 V_1}{m\langle r^2 \rangle} = \frac{1}{A^{\frac{2}{3}}}[(41)^2 + (60)^2] \text{ MeV}^2. \qquad (3.52)$$

It is seen that the interaction plays a role, in determining the energy of the dipole vibration, which is somewhat larger than that played by the kinetic energy of the nucleons. From the above relation one obtains

$$\hbar\omega_D \approx \frac{73}{A^{\frac{1}{3}}} \approx \frac{87}{R} \text{ MeV}, \qquad (3.53)$$

where $R = 1.2A^{1/3}$ is the nuclear radius measured in fm. A parametrization of the data for medium-heavy nuclei is provided by the relation (cf. Fig. 2.5(b) and Eq. (2.1))

$$\hbar\omega_D \approx \frac{95}{R} \text{ MeV}. \qquad (3.54)$$

The theoretical prediction (3.53) is close to the experimental evidence, although slightly lower.

To complete this subsection, we shall write the square energy centroid of the giant dipole resonance given in Eq. (3.52), in a more compact way which shall prove useful in the discussion of the properties of the giant vibration at finite temperature. For this purpose we recall that the energy quantum $\hbar\omega_0 \approx 41\, A^{-\frac{1}{3}}$ MeV is obtained by adjusting the frequency ω_0 to reproduce the nuclear mean-square radius. This leads to the relation

$$\sum_{k=1}^{A}\langle r_k^2\rangle = \frac{\hbar}{m\omega_0}\sum_{k=1}^{A}\left(N_k + \frac{3}{2}\right) = \frac{3}{8}AR_0^2.$$

From this condition and from the expression $A = \frac{2}{3}(N_{max}+2)^3$ valid for the harmonic oscillator one obtains

$$\hbar\omega_0 \approx \frac{5}{4}\left(\frac{3}{2}\right)^{\frac{1}{3}}\frac{\hbar^2}{mr_0^2}A^{-\frac{1}{3}} \approx \frac{41}{A^{\frac{1}{3}}}\ \text{MeV}. \qquad (3.55)$$

Here r_0 is the radius parameter defined in Eq. (3.8). From the fact that, in the Fermi gas model, the Fermi energy can be written as

$$\varepsilon_F \approx \frac{\hbar^2}{2m}\left(\frac{9\pi}{8}\right)^{\frac{2}{3}}\frac{1}{r_0^2},$$

and from the relations given in Eqs. (3.47) and (3.55), one can write

$$(\hbar\omega_0)^2 \approx \frac{1.4\hbar^2}{m\langle r^2\rangle}(b_{sym})_{kin}. \qquad (3.56)$$

To be noted is that the quantity r_0 appearing in the expression of ε_F is connected with the density radius ($r_0 = 1.12$ fm), while that appearing in Eq. (3.55) is associated with the potential radius ($r_0 = 1.2$ fm). We have written the corresponding result for $(\hbar\omega_0)^2$ in terms of this last parameter. The energy centroid of the giant dipole vibration can now be written as

$$(\hbar\omega_D)^2 = \frac{1.4\hbar^2}{m\langle r^2\rangle}(b_{sym})_{kin} + \frac{3\hbar^2}{m\langle r^2\rangle}(b_{sym})_{pot}. \qquad (3.57)$$

Because $(b_{sym})_{pot} \approx (b_{sym})_{kin} \approx \frac{1}{2}b_{sym}$, the above expression leads to

$$\hbar\omega_D \approx \sqrt{\frac{2.2\hbar^2 b_{sym}}{m\langle r^2\rangle}}. \qquad (3.58)$$

The Random Phase Approximation allows to distinguish the contributions to $\hbar\omega_D$ arising from the kinetic energy of the particles as from the isovector residual interaction, as testified by Eq. (3.52). On the other hand, the existence of a giant dipole vibration is intimately related to the fact that protons and neutrons like to be together, that is, to the isospin independence of the nuclear forces. This is reflected, among other things, in the distribution of single-particle levels generated by the mean field. In particular, in the high degeneracy ascribed by this field to particle–hole excitations which change by one the principal quantum number. It is then sensible that $(\hbar\omega_0)^2 \sim (b_{sym})_{kin}$, in the same sense that $\kappa_1 \sim (b_{sym})_{pot}$.

The result embodied in Eqs. (3.57)–(3.58) is another example of the general features displayed by many-body systems and already mentioned in the beginning of this chapter, namely, that the properties of the mean field in general and of the single-particle motion in particular, emerge from exactly the same features of the nuclear forces that are responsible for the existence of collective motion.

3.2.5 Frequency of quadrupole vibrations

To calculate the frequency of quadrupole vibrations, use shall be made of a quadrupole–quadrupole separable residual interaction (cf. also Eq. (3.41))

$$v(\vec{r}, \vec{r}') = -\kappa_2 F_{2M}(\vec{r}) F_{2M}^*(\vec{r}'), \tag{3.59}$$

with

$$F_{2M}(\vec{r}) = r^2 Y_{2M}(\hat{r}). \tag{3.60}$$

The time-dependent displacement associated with this field, making the assumption that the nucleus reacts as an incompressible fluid is

$$\vec{u}(\vec{r}, t) = \alpha(t)\vec{u}_0(\vec{r}),$$

with

$$\vec{u}_0(\vec{r}) = \vec{\nabla} F_{2M}(\vec{r}).$$

The time dependence of $\vec{u}(\vec{r}, t)$ is carried by the collective coordinate $\alpha(t)$ (cf. Eqs. (3.14)–(3.17)). The incompressibility follows from the relation

$$\vec{\nabla} \cdot \vec{u} = \alpha \nabla^2 F_{2M}(\vec{r}) = 0.$$

Being $\vec{\nabla} \times \vec{u} = 0$, the motion is also irrotational. The transition density associated with the above displacement is

$$\delta\varrho = \varrho(\vec{r} + \vec{u}) - \varrho(\vec{r}) = \alpha(t)\vec{u}_0(\vec{r}) \cdot \vec{\nabla}\varrho.$$

The transition potential must be consistent with this displacement field. Let us carry out the calculation for a generic field $F_{LM} = r^L Y_{LM}$ and then particularize it for $L = 2$, that is,

$$\delta\mathcal{U} = \vec{u} \cdot \vec{\nabla}\mathcal{U} = \alpha(t)\vec{\nabla} F_{LM} \cdot m\omega_0^2 \vec{r} = \alpha(t) m\omega_0^2 L F_{LM}. \tag{3.61}$$

In this estimate, the harmonic oscillator potential has been used to describe the static field, that is, $\mathcal{U}(r) = \frac{1}{2}m\omega_0^2 r^2$. The transition potential can also be calculated in terms of the convolution of the transition density and the two-body interaction (Eq. (3.59)), according to

$$\delta\mathcal{U} = \kappa_L F_{LM}(\vec{r}) \int d^3r' \, F_{LM}^*(\vec{r}')\delta\varrho. \tag{3.62}$$

Equating the results of Eqs. (3.61) and (3.62) one obtains

$$\kappa_L = -\frac{L\alpha(t)m\omega_0^2}{\int d^3r' F_{LM}^*(\vec{r'})\delta\varrho}. \tag{3.63}$$

The integral in this equation is given by

$$
\begin{aligned}
\alpha(t)\int d^3r' F_{LM}^* \vec{\nabla} F \cdot \vec{\nabla}\varrho &= -\alpha(t)\int d^3r' |\nabla F|^2 \varrho \\
&= -L(2L+1)\alpha(t)\int dr\, r^{2L-2}\varrho \\
&= -L(2L+1)\alpha(t)\frac{A}{4\pi}\langle r^{2L-2}\rangle, \tag{3.64}
\end{aligned}
$$

where Eq. (3.44) has been used. Inserting the result of Eq. (3.63) one obtains

$$\kappa_L = \frac{4\pi m\omega_0^2}{(2L+1)A\langle r^{2L-2}\rangle}. \tag{3.65}$$

For $L = 2$ the result is

$$\kappa_2 = \frac{4\pi m\omega_0^2}{5\langle r^2\rangle}. \tag{3.66}$$

Giant quadrupole excitations are produced by promoting particles from an occupied shell with principal quantum number N (harmonic oscillator) to unoccupied shells with principal quantum numbers $N + 2$, $N + 4$ etc. This in keeping with the fact that the parity of the single-particle states is $(-1)^N$. Because in the harmonic oscillator the only non-diagonal matrix elements of the field $r^2 Y_{2M}$ are

$$\langle N'|r^2|N\rangle \propto \delta(N', N \pm 2),$$

the particle–hole excitation energy associated with quadrupole modes is $\varepsilon_{\nu_k} - \varepsilon_{\nu_i} = 2\hbar\omega_0$. The dispersion relation given in Eq. (3.30) can be written as

$$\frac{\sum_{\nu_k,\nu_i} 2(\varepsilon_{\nu_k} - \varepsilon_{\nu_i})|\langle \tilde{\nu}_i|r^2 Y_2|\nu_k\rangle|^2}{(2\hbar\omega_0)^2 - (\hbar\omega_Q)^2} = \frac{1}{\kappa_2}. \tag{3.67}$$

Making use of the quadrupole energy weighted sum rule (cf. Eq. (3.38))

$$\sum_{\nu_k,\nu_i}(\varepsilon_{\nu_k} - \varepsilon_{\nu_i})|\langle \tilde{\nu}_i|r^2 Y_2|\nu_k\rangle|^2 = \frac{15}{4\pi}\frac{\hbar^2}{m}AR_0^2, \tag{3.68}$$

one obtains

$$\hbar\omega_Q = \sqrt{(2\hbar\omega_0)^2 - 2(\hbar\omega_0)} = \sqrt{2}\hbar\omega_0 = \frac{58}{A^{\frac{1}{3}}} \text{ MeV}. \tag{3.69}$$

In keeping with the fact that the mode under consideration is isoscalar, the bare mass appears in Eq. (3.68) (cf. discussion following Eq. (3.37)), and thus the result given in Eq. (3.69) is not modified by the energy dependence of the single-particle potential.

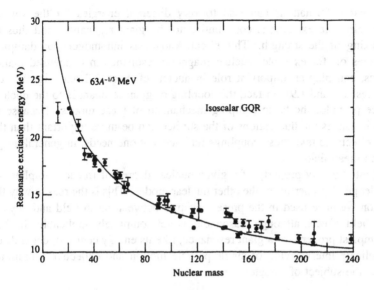

Figure 3.8.   Energy systematics of the giant quadrupole resonance.

In Fig. 3.8 we display the systematics of centroids of the giant quadrupole as a function of mass number. The results are well parametrized by the function

$$\hbar\omega_Q \approx \frac{63}{A^{\frac{1}{3}}} \text{ MeV}. \tag{3.70}$$

3.3   Damping of Nuclear Motion

It is satisfying that mean field theory provides an overall account of the energy centroids of giant resonances. However, mean field theory predicts essentially sharp states, while the observed resonances display conspicuous widths Γ. Typically, $\hbar\omega_{GR}/\Gamma \sim 3\text{--}5$, implying that the vibration goes through few periods of oscillation before its energy and angular momentum either leaves the nucleus or is distributed among the nucleons (cf. Fig. 2.5(c)).

As already mentioned in Chapter 1, at the level of mean field, there are four mechanisms which can damp the vibrational motion: (a) decay into single-particle motion, (b) deformation effects, (c) particle-decay, (d) γ-decay.

The decay of the vibration into single-particle motion, known in the damping of plasmons in infinite systems as Landau damping, plays little role in the nuclear case, exception made in light nuclei. This in keeping with the fact that the centroid of giant resonances is found at energies corresponding approximately to those typical of uncorrelated particle–hole excitation of opposite parity (the positive parity GQR centroid $\hbar\omega_Q$ is close in energy to negative-parity particle–hole excitation of energy $\hbar\omega_0 \approx 41 A^{-1/3}$ MeV, while $\hbar\omega_D$ almost coincides with positive parity particle-hole excitation of energy $2\hbar\omega_0$).

Nuclear deformations can lead to very different energies for the variety of particle–hole excitations contributing to the giant resonance, and thus to a breaking of the strength. This effect, known as inhomogeneous damping in studies of, for example, nuclear magnetic resonance in condensed matter physics, can play an important role in nuclei (cf. Sect. 2.2 and Fig. 2.5, cf. also Figs. 1.8 and 1.9). In fact, the coupling of giant resonances to the nuclear surface provides the basic damping mechanism of these modes. Because in many instances the fluctuations of the surface can be more important than the static effects, to take these couplings into account one needs, in general, to go beyond mean field.

Essentially, any property of a given nuclear degree of freedom implies the knowledge of properties of the other nuclear modes. This is the reason why the division we have used in the present chapter between mean field and beyond mean field effects, although reasonable, is not completely applicable. In fact, the simplest property of a giant resonance, like its energy centroid, depends on the delicate interweaving of single-particle motion and collective vibrations. This is the subject of Chapter 4.

4

Beyond Mean Field

The mechanism for a nucleon to change its state of motion from a single-particle orbital to a different one, is by colliding with another nucleon, promoting it from a state in the Fermi sea to a state above the Fermi surface, as shown in Fig. 4.1. Because of the Pauli principle, processes of this type are more probable to occur at the nuclear surface. This region of the nucleus plays a central role in the case of both density and pairing fluctuations. In keeping with the fact that, in the nuclear case, particle–particle correlations are considerably weaker than particle–hole correlations, the collision process depicted in Fig. 4.1 is well described by the coupling of the nucleon to a collective surface vibration (cf. Fig. 3.3). This is the doorway phenomenon, the states containing a nucleon and a vibration being the doorway states. The concept of doorway state was introduced in the work of Block and Feshbach (1963). The general formulation of the question was worked out in Feshbach et al. (1967) (cf. also Sect. 4.3).

4.1 Doorway States

Through the coupling introduced in Eq. (3.16) a particle can set the nuclear surface into vibration. Such process can be repeated, the particle interacting a second time with the surface and reabsorbing the vibration (cf. Fig. 4.2). Thus, the properties characterizing the particle like mass, charge, mean free path, occupation number, etc., can be modified.

4.1.1 The dynamical shell model

The scattering processes depicted in Fig. 4.2 are some of the most relevant processes in the particle-vibration coupling renormalizing the single-particle

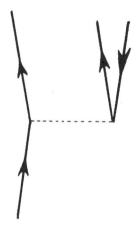

Figure 4.1. Collision between nucleons where a particle changes state of motion by inducing a particle–hole excitation.

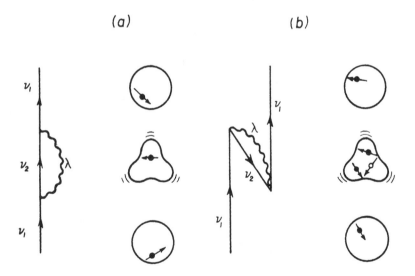

Figure 4.2. The lowest-order process by which the single-particle motion is renormalized by the coupling to the nuclear surface. In (a) the particle excites the vibration by bouncing inelastically off from the surface. In (b) the vibration is excited by a virtual process (vacuum fluctuation). Particles are represented by an upwardgoing arrowed line (by a solid dot) while holes are pictured as a downwardgoing arrowed line (open circle). The surface vibration is drawn as a wavy line.

motion. The associated expression of the self-energy operator is

$$\Sigma(1,\omega) = \sum_{2,\lambda} \frac{V^2(1,2;\lambda)}{\omega - (\varepsilon_2 + \omega_\lambda)} + \frac{V^2(1,2;\lambda)}{\omega - (\varepsilon_2 - \omega_\lambda)}. \tag{4.1}$$

Each term of this expression has the typical structure of an energy correction in second order perturbation theory, that is, a square matrix element divided by an energy denominator. The energy denominators are given by the difference in energy between the initial and the intermediate states. The energy of particle 1 has been denoted ω and that of particle 2, ε_2. The energy of the vibration is labeled by ω_λ. The quantities $V(1,2;\lambda)$ are the matrix elements coupling the particle to the vibration (cf. Eq. (3.22)).

Because the energy of the intermediate state may coincide with the energy ω of the single-particle, one has to replace ω by $\omega + i\frac{I}{2}$, where I represents the energy interval over which averages are carried out. The self-energy operator can then be written as

$$\Sigma(1, \omega + iI) = \Delta E(1, \omega + iI) - \frac{i}{2}\Gamma(1, \omega + iI), \tag{4.2}$$

sum of a real and of an imaginary term. The final result, obtained by taking the limit of $\Sigma(1, \omega + iI)$ for $I \to 0$, should not depend on the averaging process.

It is illuminating to calculate the imaginary part of the self-energy, that is,

$$\Gamma(1,\omega) = \sum_{2,\lambda} V^2(1,2;\lambda) \frac{I}{(\omega - (\varepsilon_2 + \omega_\lambda))^2 + (\frac{I}{2})^2}. \tag{4.3}$$

Taking the limit of this function for $I \to 0$ one obtains

$$\Gamma(1,\omega) = 2\pi \sum_{2,\lambda} V^2(1,2;\lambda)\delta(\omega - (\varepsilon_2 + \omega_\lambda)). \tag{4.4}$$

Approximating $V(1,2;\lambda)$ by its average value V leads to

$$\Gamma(1,\omega) = 2\pi V^2 \varrho(\omega), \tag{4.5}$$

where the quantity $\varrho(\omega) = \sum_{2,\lambda} \delta(\omega - (\varepsilon_2 + \omega_\lambda))$ is the density of final states per unit energy, into which the particle state can decay. The above relation, known as Fermi's Golden Rule, is the basic expression used to describe the decay width of a quantal state. For scattering states, the quantity $-\frac{1}{2}\Gamma$ can be identified with the imaginary part of the optical potential (cf. Sect. 4.1.2).

There is extensive experimental evidence, which testifies to the fact that a nucleon moving in an orbital close to the Fermi energy has a mean free path which is large compared with the nuclear dimensions. Consequently, it being in a stationary state, its wave function can be written as $\varphi_1(\vec{r}, t) = \varphi_1(\vec{r})e^{-i\omega t}$. For single-particle levels progressively removed from the Fermi energy, the probability of finding states with the same energy of the single-particle state, in particular states built out of a single-particle and a collective surface vibration, becomes sizable. Under these circumstances, the single-particle levels acquire a width (cf. Eq. (4.4)). The associated wave function can be written as $\varphi_1(\vec{r},t) = \varphi_1(\vec{r})e^{-i\omega t}e^{-\frac{\Gamma}{2\hbar}t}$. Consequently, the probability of finding the state 1 occupied by a particle at time t, when it was occupied with probability 1 at time $t = 0$ is given by $\int d^3r \, |\varphi_1(\vec{r},t)|^2 = \exp^{-\frac{\Gamma}{\hbar}t}$. The associated lifetime can be written as

$$\tau = \frac{\hbar}{\Gamma}, \tag{4.6}$$

following Heisenberg's uncertainty relation.

The mean-free path is defined as the decay length associated with a stationary state in time. For a wavefunction of the type $\exp(ik_R r - k_I r)$, the mean-free path is $\lambda = 1/2k_I$, where k_R and k_I are the real and imaginary components of the single-particle linear momentum. Requiring that the expectation value of the Hamiltonian $H_{s.p.}$ is real and making use of the fact that single-particle kinetic energy is $\hbar^2 k^2/2m_k$ (cf. Sect. 3.1.1), one obtains $(\hbar^2/m_k)k_R k_I = \Gamma/2$. Thus

$$\lambda = v_g \tau = \frac{\hbar^2 k_F}{m_k \Gamma}. \tag{4.7}$$

Here v_g is equal to the group velocity of the particle k_R/m_k.

The coupling of the single-particle motion to surface vibrations can, in principle, only lead to a breaking of the single-particle strength and not to a damping process. However, these processes, known as "doorway coupling"(Block and Feshbach (1963); cf. also Feshbach et al. (1967)), are the first in a hierarchy of couplings to progressively more complicated states which eventually lead to the coupling of the single-particle levels to the compound nucleus (cf. Figs. 2.8 and 4.18). In this process, the energy and angular momentum of the single-particle state is uniformly distributed among all nucleons. This is the phenomenon of single-particle damping.

It is well established that the coupling of the single-particle motion to "doorway states" spreads the single-particle states over an energy range which is of the same order of magnitude as that experimentally observed (cf., e.g., Bertsch et al. (1983)). The coupling to more complicated configurations essentially does not alter this result (cf. Sect. 4.2.2). Within the formalism discussed above, the small parameter I represents, in some average way, the more complicated couplings. For scattering states, that is a particle in the continuum, the imaginary part of the optical potential $W(= -\frac{1}{2}\Gamma)$ leads to a decrease of the incident flux describing all non-elastic processes.

Before proceeding further, let us briefly recall some properties of analytic functions which may prove of use within the context of the present discussion. Given a function $f(z)$ in the complex plane which displays a first order pole at $x = x_0$, Cauchy's principal part integral is defined as

$$\mathcal{P}\int_a^b dz \, f(z) = \lim_{\epsilon \to 0+} \int_a^{x_0-\epsilon} dz \, f(z) + \lim_{\epsilon \to 0+} \int_{x_0+\epsilon}^b dz \, f(z).$$

One can always define $f(z) \equiv g(z)/(z-z_0)$, where $g(z)$ is analytic throughout and $g(x_0) \neq 0$. One can then write

$$\mathcal{P} \int dz \, \frac{g(z)}{z-z_0} = i\pi g(z_0).$$

Identifying $g(z)$ with $\Sigma(1, \omega + iI)$ and taking the imaginary part of Eq. (4.2), one obtains

$$\Delta E(1; z) = \frac{\mathcal{P}}{\pi} \int dz' \, \frac{-\frac{1}{2}\Gamma(1; z')}{z' - z} \tag{4.8}$$

The knowledge of $\Gamma(1; \omega + iI)$ for all values ω completely defines $\Delta E(1, \omega + iI)$. This is a very useful relation. Because processes in which the particle undergoes a real transition (on-the-energy-shell processes) that is, transitions where the energy of the intermediate state coincides with that of the particle, are easier to calculate than virtual processes (off-the-energy-shell processes), where the intermediate state can have any energy value. Aside from providing an economic way to compute the real part of the self-energy, Eq. (4.8) is a particular form of the so called Kramers-Krönig dispersion relations embodying the condition of causality to be fulfilled by the behaviour of any quantal system, in particular that of a nucleon in the nucleus (cf., e.g., Mahan (1981)).

From the knowledge of the real and imaginary parts of the single-particle self-energy one can construct the single-particle strength function,

$$P_1(\omega) = \frac{1}{2\pi} \frac{\Gamma + I}{(\varepsilon_1 + \Delta E - \omega)^2 + \frac{1}{4}(\Gamma + I)^2}. \tag{4.9}$$

This function is normalized according to $\int d\omega \, P_1(\omega) = 1$. In other words, the expression given in Eq. (4.9) describes how the single-particle strength originally concentrated in the shell-model state 1 with energy ε_1, is distributed over more complicated states composed essentially of a particle and a surface vibration, as a function of the excitation energy.

The properties characterizing the strength function in a given energy interval $\omega_i \leq \omega \leq \omega_f$ are the centroid

$$\langle \omega \rangle = \frac{\int_{\omega_i}^{\omega_f} d\omega \, P_1(\omega)\omega}{\int_{\omega_i}^{\omega_f} d\omega \, P_1(\omega)}, \tag{4.10}$$

and the different central multiple moments,

$$\mu_n = \langle (\omega - \langle \omega \rangle)^n \rangle. \tag{4.11}$$

The average quantity A is defined through the expression

$$\langle A \rangle = \frac{\int_{\omega_i}^{\omega_f} d\omega \, P_1(\omega)A}{\int_{\omega_i}^{\omega_f} d\omega \, P_1(\omega)}. \tag{4.12}$$

The quantity

$$P_1 = \int_{\omega_i}^{\omega_f} d\omega \, P_1(\omega), \tag{4.13}$$

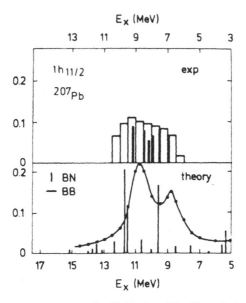

Figure 4.3. The upper part represents the distribution of the $1h_{\frac{11}{2}}$ single-particle strength as measured between 6.7 and 10.5 MeV excitation energy in ^{207}Pb by the ^{208}Pb(^{3}He,α) pick-up experiments reported in J. Guillot et al. (1980) and S. Galès et al. (1978). The measured excitation energy is shown on the upper horizontal scale. The lower part gives the spectroscopic factors (vertical segments) as calculated in Nguyen van Giai (1980) and the spectral functions $1h_{11/2}$ (full curve with dots) as calculated in Bortignon and Broglia (1981), versus the calculated excitation energy given by the lower horizontal scale. (Taken from S. Galès (1980).)

gives the probability of finding the state 1 in the energy interval $\omega_i \leq \omega \leq \omega_f$. Instead of the second moment it is customary to use the standard deviation

$$\sigma = (\mu_2)^{\frac{1}{2}}, \tag{4.14}$$

and instead of the third moment, the skewness

$$s = \mu_3 / (\mu_2)^{\frac{1}{2}}. \tag{4.15}$$

In the case of a Gaussian strength function

$$\Gamma = \sqrt{8ln2}\,\sigma \approx 2.4\sigma. \tag{4.16}$$

Examples of the strength function for single-hole states in ^{207}Pb and of a single quasi-particle state in ^{119}Sn are displayed in Figs. 4.3 and 4.4 and in Table 4.1.

Although much is yet to be learned about the damping of single-particle states, a simple parametrization of the corresponding damping width is provided by the relation (cf. Fig. 4.5)

$$\Gamma^{\downarrow}_{sp} \approx 0.5\omega, \tag{4.17}$$

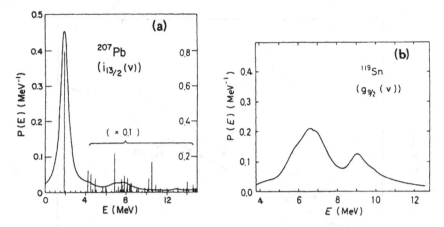

Figure 4.4. Strength function (spectral distribution) associated with (a) the $i_{13/2}$ neutron hole states of ^{207}Pb and (b) the neutron quasiparticle state in ^{119}Sn. The averaging parameter I was chosen equal to 1.0 MeV. For the case of the $i_{13/2}^{-1}(\nu)$ hole state the results obtained setting $I = 0$ are also shown. All lines, with the exception of that at $\sim$ 2 MeV, were multiplied by a factor of 10. Of the two scales, the one to the left is associated with the calculations carried out with $I = 1$ MeV, while that to the right corresponds to $I = 0$ and gives the spectroscopic factor $C^2 S/(2j + 1)$. (From Bortignon and Broglia (1981).)

Table 4.1. The centroid, single-particle spectroscopic factors and standard deviation (cf. Eqs. (4.10)–(4.14)) of the strength function associated with the neutron hole state $i_{13/2}$ of ^{207}Pb and with the neutron state of ^{119}Sn with quantum numbers $1g_{9/2}$ (cf. Fig. 4.4), calculated (Bortignon and Broglia (1981)) within the energy range 4.3 MeV $< E <$ 6.5 MeV as experimentally measured (cf. Galès et al. (1978), Guillot et al. (1980)).

nlj	$(^A X)$	$\langle \varepsilon_i \rangle$ (MeV)		$C^2 S$		σ (MeV)	
		theory	exp.	theory	exp.	theory	exp.
$1i_{13/2}$	$(^{207}$Pb)	2.04	2.30	11.5	13.7	0.62	1.5
$1g_{9/2}$	$(^{119}$Sn)	5.6	5.9	2.5	2.5	0.45	0.49

where ω is the single-particle energy measured from the Fermi energy ($\omega = E - \varepsilon_F$). This parametrization is supported by detailed calculations, Bertsch et al. (1979), Bortignon et al. (1986), Donati et al. (1996) (cf. Sect. 9.1.1).

To emphasize the fact that the relation given in Eq. (4.17) is the damping width arising from the coupling of single-particle states with more complicated configurations, we have written an arrow pointing downwards as superscript. A similar result to that displayed in Eq. (4.17) is obtained from the imaginary part of the optical potential. From the above relation and Eq. (4.7) one can write

$$\lambda \approx \frac{150 \text{ fm}}{\omega}, \tag{4.18}$$

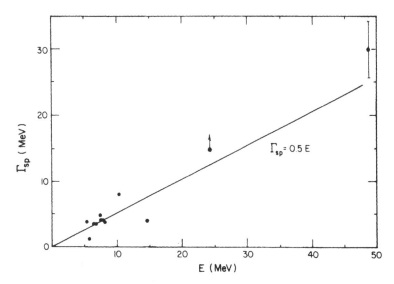

Figure 4.5. Full width at half maximum of the strength function associated with deep hole states, bound states, and scattering states in a variety of nuclei. (From Bertsch et al. (1983).)

where the energy ω is measured in MeV. In terms of the nuclear diameter $d = 2R \approx 2.4A^{\frac{1}{3}}$ fm, we find that, for medium-heavy nuclei ($100 \leq A \leq 200$) $\lambda/d \approx 15/\omega$.

4.1.2 Effective mass (ω-mass)

As the graphical perturbation expansion of the single-particle self-energy suggests (cf. Fig. 4.2 and Eq. (4.2)), the Hamiltonian describing the single-particle motion reads (cf. also Eq. (3.10)),

$$H_{s.p.} = (-\frac{\hbar^2}{2m}\nabla^2 + \tilde{V}(k) + \Delta E(\omega)) + \tilde{U}(r) + iW(\omega), \qquad (4.19)$$

where $W = -\frac{1}{2}\Gamma$. The dependence on the momentum of the particle is associated with the non-locality arising from the Pauli principle, and has been discussed in Section 3.1. The dependence on the frequency of the particle is associated with the non-locality in time generated by the coupling to a surface vibration excited by the particle at a given time and reabsorbed at a different time.

For many purposes it is possible to rewrite the term in parenthesis as a kinetic energy term with an effective mass m^* (cf., e.g., Mahaux et al. (1985)). In fact, requiring that

$$\frac{d\hbar\omega}{dk} = \frac{\hbar^2 k}{m^*}, \qquad (4.20)$$

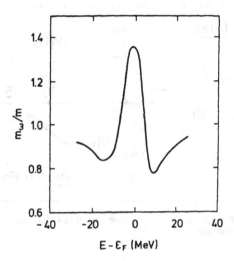

Figure 4.6. The ratio m_ω/m of the ω-mass of a nucleon in ^{208}Pb to the bare nucleon mass as a function of the energy of the particle measured with respect to the Fermi energy, calculated within the particle–vibration coupling model. (After Mahaux et al. (1985).)

and calculating (cf. Eq. (4.19))

$$\frac{d\hbar\omega}{dk} = \frac{\hbar^2 k}{m} + \frac{\partial\tilde{V}(k)}{\partial k} + \frac{\partial\Delta E(\omega)}{\partial\hbar\omega}\frac{d\hbar\omega}{dk}, \qquad (4.21)$$

which is equivalent to

$$\frac{d\hbar\omega}{dk} = \frac{\hbar^2 k}{m}\left(1 - \frac{\partial\Delta E}{\partial\hbar\omega}\right)\left(1 + \frac{m}{\hbar^2 k}\frac{\partial\tilde{V}(k)}{\partial k}\right)^{-1}, \qquad (4.22)$$

one obtains

$$\frac{m^*}{m} = \frac{m_k}{m}\frac{m_\omega}{m}. \qquad (4.23)$$

In this equation the ω-mass is given by

$$\frac{m_\omega}{m} = \left(1 - \frac{\partial\Delta E(\omega)}{\partial\hbar\omega}\right) \qquad (4.24)$$

while m_k/m coincides with the k-mass defined in Sect. 3.1.1. Consequently,

$$H_{s.p.} = -\frac{\hbar^2}{2m^*}\nabla^2 + \tilde{U} + iW, \qquad (4.25)$$

which is the optical-model Hamiltonian.

In Fig. 4.6 we display results of calculations of the ω-mass for the single-particle and single-hole states of ^{208}Pb. The quantity $\frac{m_\omega}{m}$ displays a peak as a function of the single-particle energy centered around ε_F. The associated FWHM is ≈ 10 MeV, that is, the ω-mass increase over the bare mass happens

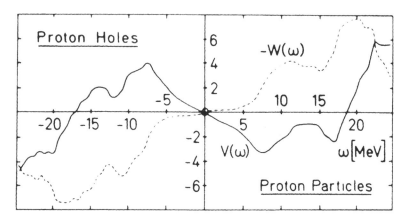

Figure 4.7. Dependence upon $\omega = E - \varepsilon_F$ of the quantities $W(\omega)$ and $\Delta E(\omega)$ defined in the text. (Taken from Sommerman et al. (1983).)

in the interval of energy comprised between -5 MeV to $+5$ MeV around the Fermi energy. This width is controlled by the frequency of collective surface vibrations which, in ^{208}Pb, have an average energy of the order of 3.3 MeV ($\hbar\omega_{3^-} = 2.6$ MeV, $\hbar\omega_{5^-} = 3.2$ MeV, $\hbar\omega_{2^+} = 4.1$ MeV). Consequently, for frequencies of the single-particle much higher than this value, the phonons cannot dress the particle in an efficient way any more. Within the same interval of energy around the Fermi energy for which $m_\omega/m > 1$, the imaginary part of the self-energy (cf. Figs. 4.7 and 4.8) is essentially zero. This is because no real transition exists in this energy interval.

From these results one can understand why the empirical evidence concerning the energy of single-particle levels around the Fermi energy is well described by the motion of nucleons in a real, energy independent, average potential, carrying a mass equal to the bare nucleon mass. Furthermore, the result that the ω-mass is larger than the bare mass has the consequence that the density of levels around the Fermi energy is larger than that predicted by Hartree–Fock theory, in accordance with the experimental findings (cf. Fig. 3.2). However, there is a basic difference between this simple model and the results expressed by Eq. (4.25). In fact, in the empirical shell model (independent-particle model), the occupation of each level is either 1 or 0. The situation is more subtle here. In fact, due to its coupling to the nuclear surface, a particle which starts in a pure single-particle configuration is forced into more complicated states of motion. Consequently, the probability of finding a particle in a single-particle state below the Fermi level is different from 1. Similarly, unoccupied states at the level of the pure shell model become partially occupied as the particle jumps to these states by exciting a surface mode.

The fact that the "more complicated" states to which the particle states couple can be at a higher energy than the original energy available to the

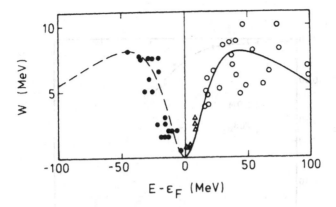

Figure 4.8. Dependence upon $\omega = E - \varepsilon_F$ of the strength of the imaginary part of the optical-model potential for nuclei with mass number $12 \leq A \leq 60$. The full dots are obtained from the spreading widths compiled in Jacob and Maris (1973). The triangles are derived from neutron strength functions of Cohen (1965), Cohen (1973). The open dots are deduced from the analysis of elastic cross sections (Giannini et al. (1980). (Taken from Mahaux and Ngô (1981)).)

particle presents no contradiction, as these are virtual states, that is, states which last a finite amount of time. Because of Heisenberg's relations energy does not need to be conserved within a range which becomes larger the shorter the time the intermediate state is virtually excited. However, an external field, as that produced by a proton can provide the necessary energy to make the process real and eventually pick-up a neutron in the reaction $A(p, d)B$ from states above the Fermi energy. Results of calculations of the occupation number

$$
n_j = \begin{cases} 1 + \frac{d\Delta E'}{dE} & j= \text{occ. orbit,} \\[3mm] -\frac{d\Delta E'}{dE} & j= \text{empty orbit,} \end{cases} \tag{4.26}
$$

are given in Fig. 4.9. In the above equation, the quantity $\Delta E'$ is the contribution associated with the diagram of Fig. 4.2 which arises from ground state correlations.

4.2 Relaxation of Giant Vibrations

One can view giant vibrations as an excitation built out of a particle above the Fermi surface and of a hole in the Fermi sea. A first estimate of the damping width of giant vibrations can be obtained assuming that the particle and the hole couple to more complicated configurations acquiring a width. Then the total width is the sum of individual widths. Because in the damping process we deal with real processes, that is processes where the energy is conserved, the energy of the resonance has to be shared between the particle and the hole.

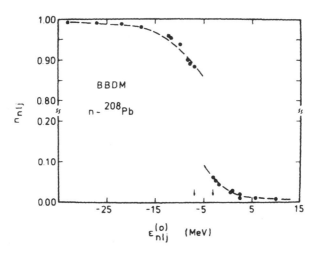

Figure 4.9. Occupation probability of neutron orbits in the correlated ^{208}Pb nucleus plotted versus the single-particle energy ε_{nlj} computed in the Skyrme III–Hartree–Fock approximation. The calculation is based on Eq. (4.26). The dots correspond to the $1f_{7/2}, 2p_{1/2}, 1g_{7/2}, 1h_{11/2}, 1h_{9/2}, 2f_{7/2}, 2f_{5/2}, 1i_{13/2}, 3p_{3/2}$ and $3p_{1/2}$ hole states and to the $2g_{9/2}, 1i_{11/2}, 1j_{15/2}, 3d_{5/2}, 2g_{7/2}, 4s_{1/2}, 3d_{3/2}, 2h_{11/2}$ and $2h_{9/2}$ particle states. The dashed curve has been drawn to guide the eye through the calculated dots in order to exhibit their trend. The arrows show the location of $\varepsilon_F^- = \varepsilon_{3p_{1/2}}$ and $\varepsilon_F^+ = \varepsilon_{2g_{9/2}}$. (After Mahaux et al. (1985).)

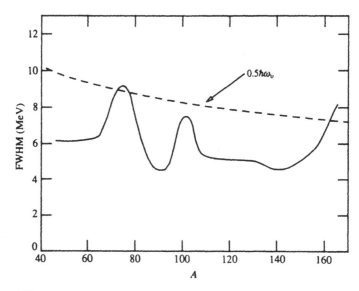

Figure 4.10. Full width half maximum (FWHM) of giant dipole resonances (Snover (1986)). The maxima correspond to the case of deformed nuclei. The dashed line corresponds to the theoretical estimate given in Eq. (4.28).

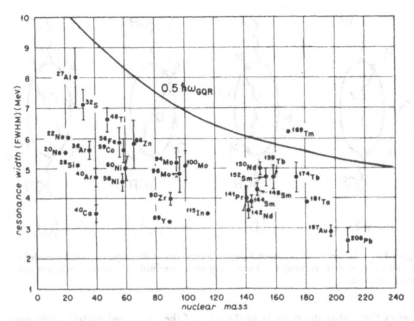

Figure 4.11. Damping width of the giant quadrupole vibration (cf. Satchler (1977)). The continuous curve corresponds to the estimate given in Eq. (4.29).

The simplest expression one can write for the giant resonance damping width is then

$$\Gamma_{GR}^{\downarrow}(\hbar\omega_{GR}) = \Gamma_p^{\downarrow}(\frac{\hbar\omega_{GR}}{2}) + \Gamma_h^{\downarrow}(\frac{\hbar\omega_{GR}}{2})$$
$$\approx 0.5\hbar\omega_{GR}, \tag{4.27}$$

where the expressions for $\Gamma_p^{\downarrow}$ and $\Gamma_h^{\downarrow}$ given in Eq. (4.17) have been used.

Making use of the expressions $\hbar\omega_0 \approx 79A^{-1/3}$ MeV (cf. Eq. (2.1)) and $\hbar\omega_Q \approx 63A^{-1/3}$ MeV (cf. Eq. (3.70)), the above equation leads to

$$\Gamma_D^{\downarrow} \approx \frac{40}{A^{\frac{1}{3}}} \text{ MeV}, \tag{4.28}$$

and

$$\Gamma_Q^{\downarrow} \approx \frac{30}{A^{\frac{1}{3}}} \text{ MeV}, \tag{4.29}$$

for the damping width of the giant dipole and quadrupole resonances, respectively. These expressions are displayed in Figs. 4.10 and 4.11 in comparison with the experimental findings. Because the expression (4.28) is valid for spherical nuclei, one can conclude that the simple estimates overpredict the experimental findings by roughly 50%.

We shall show below that the relation given by Eq. (4.27) neglects important correlation effects between the particle and the hole. In fact, this relation

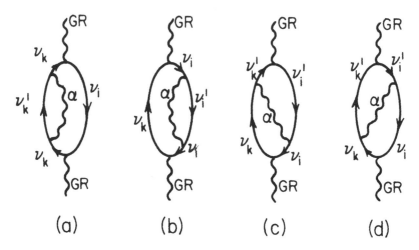

Figure 4.12. Lowest-order processes by which a resonance (GR) couples to a two particle-two hole intermediate state (doorway state) containing an uncorrelated particle–hole excitation and a surface vibration.

implies that either the particle or the hole of the correlated particle–hole pair which constitutes a resonance not only can excite a surface vibration, which is true, but have also to reabsorb the phonon they have excited. This of course is not correct, in that a surface vibration excited by the inelastic scattering of particle off the nuclear surface, can be absorbed at a later time by the hole, and vice versa. In other words, the expression (4.27) takes care only of the processes (a) and (b) of Fig. 4.12. We shall see that processes (c) and (d), where a phonon is exchanged between the fermions act as a glue between the particle and the hole preventing, to a large extent, the decay of the resonance, and reducing the contributions (a) and (b) to the damping width [1]. In fact, the self-energy correction to the giant vibration implied by the process of Fig. 4.12(a) is

$$
\begin{aligned}
\Sigma^{p}_{self-en}(GR,\omega) &= \sum_{\nu_k,\nu_i,\nu_{k'},\lambda} X^2_{GR}(\nu_k,\nu_i)\frac{V^2(\nu_k,\nu_{k'};\lambda)}{\hbar\omega - ((e_{\nu_{k'}} - e_{\nu_i}) + \hbar\omega_\lambda)} \\
&= \sum_{\nu_k,\nu_i} X^2_{GR}(\nu_k,\nu_i)\Sigma(\nu_k,\omega + e_{\nu_i}).
\end{aligned}
\tag{4.30}
$$

That is, it is the sum of the contributions of the self-energy of each particle participating in the linear combination of particle–hole excitations describing the resonance. In other words, it is the weighted average of the single-particle self-energies of all the particle–hole configurations. The weighting factor is the

---

[1] The contributions arising from graphs (c) and (d) of Fig. 4.12, together with those corresponding to graphs (a) and (b) of the same figure, are required on the basis of very general conservation rules, which can also be derived from the so called Ward identities (cf., e.g., Weinberg (1995)).

probability for the giant vibration to be in a given configuration. The particle self-energies are calculated at an energy $(\hbar\omega + e_{\nu_i})$, that is, at an energy lower than the energy of the giant resonance by the amount e_{ν_i} $(e_{\nu_i} = \varepsilon_{\nu_i} - \varepsilon_F < 0)$, which is the energy taken up by the hole of the different particle–hole excitations. A similar expression is obtained for the decay of the hole (cf. Fig. 4.12(b)), that is,

$$\Sigma^h_{self-en}(GR,\omega) = \sum_{\nu_k,\nu_i} X^2_{GR}(\nu_k,\nu_i)\Sigma(\nu_i,\omega - e_{\nu_k}) \tag{4.31}$$

where now $e_{\nu_k} = \varepsilon_{\nu_k} - \varepsilon_F > 0$.

Making the ansatz that: (a) the giant resonance is a very correlated state such that one can approximate the amplitudes by $|X| \sim \frac{1}{\sqrt{N}}$, N being the dimension of the particle–hole basis where the RPA solution of the giant vibration has been calculated, and (b) the particle-vibration coupling matrix elements are independent of the configuration, one can write, for both of the expressions given in Eqs. (4.30) and (4.31),

$$\Sigma^\nu_{self-en}(GR,\omega) \approx \Sigma(\nu,\omega - |e_{\nu'}|), \tag{4.32}$$

where ν is either a particle or a hole and ν' a hole or a particle respectively. The imaginary part of the above equation leads to the relation (4.27).

The self-energy associated with the process (d) of Fig. 4.12 is

$$\Sigma_{vertex} \quad (GR,\omega)$$
$$= \sum_{\nu_k,\nu_i,\nu_{k'},\nu_{i'}} X_{GR}(\nu_k\nu_i)X_{GR}(\nu_{k'}\nu_{i'})$$
$$\frac{\langle\nu_{i'}^{-1}|\hat{F}|\nu_i^{-1}\rangle\langle\nu_{k'}|\hat{F}|\nu_k\rangle^2\Lambda_\lambda^2}{\hbar\omega - ((e_{\nu_{k'}} - e_{\nu_i}) + \hbar\omega_\lambda)}, \tag{4.33}$$

where $|\nu_i^{-1}\rangle$ represents a state of a hole while $|\nu_i\rangle$ that of a particle moving in the same single-particle state. The matrix elements between hole states are related to those between particle states according to

$$\langle\nu_{i'}^{-1}|\hat{F}|\nu_i^{-1}\rangle = c\langle\nu_{i'}|\hat{F}|\nu_i\rangle, \tag{4.34}$$

where c is a phase (i.e., $c^2 = 1$) defined through the relation

$$(\tau\hat{F}\tau^{-1})^\dagger = -c\hat{F}. \tag{4.35}$$

Here τ stands for the time reversal operator and the dagger identifies hermitian conjugation. Because in average $\langle\nu_{i'}|F|\nu_i\rangle$ and $\langle\nu_{k'}|F|\nu_k\rangle$ have the same order of magnitude one can approximate the last expression by

$$\Sigma_{vertex}(GR,\omega) \approx$$
$$c\sum_{\nu_k,\nu_i,\nu_{k'}} (X_{GR}(\nu_k\nu_i))^2 \frac{V^2(\nu_{k'},\nu_k;\lambda)}{\hbar\omega - ((e_{\nu_{k'}} - e_{\nu_i}) + \hbar\omega_\lambda)}. \tag{4.36}$$

A similar expression is obtained for the process depicted in Fig. 4.12(c). Consequently,

$$\frac{\Sigma_{vertex}}{\Sigma_{self-en}} \approx c. \tag{4.37}$$

Because the single particle field $\hat{F}$ is a spin–isospin independent field (cf. Eq. (3.21)), $c = -1$. The physical reason of the minus sign in the phase relating processes (a) and (d) of Fig. 4.12 is associated with the fact that the multipole moments of a particle and of a hole have different sign, in keeping with the fact that closed shells systems are spherical.

Under the approximation leading to Eq. (4.36), there would be a complete cancellation between the different processes contributing to the self-energy operator of the giant resonance, and eventually to its damping width. The fact that the subspaces available to the particles (ν_k) and to the holes (ν_i) are different, makes the approximations used above not quantitatively accurate although they are qualitatively correct. The cancellation implied by Eq. (4.37), although being conspicuous, is not complete.

Numerical calculations indicate that the cancellation discussed above implies a reduction of the contributions stemming from particle- and hole-decay of the order of 30–50%, bringing theory into overall agreement with the experimental findings. An example of such calculations is discussed in the next subsection.

4.2.1 Giant quadrupole resonance

In the example to be discussed below, the giant quadrupole resonance of ^{208}Pb, the mean field solutions were obtained solving the Hartree–Fock equations (3.2) with an effective nucleon–nucleon interaction known as Skyrme III.

Making use of the resulting single-particle levels and wavefunctions, a particle–hole basis was constructed for multipolarities in the range $0 \leq L \leq 12$. The upper limit of L was fixed in order to satisfy the basic ansatz made in formulating the theory of vibrations in nuclei (cf. Sect. 3.2), that is, that the system can be viewed as a continuous medium. Consequently, the wavelength of the mode cannot be shorter than the average distance between nucleons (Debye cut-off),

$$d \approx \left(\frac{\frac{4\pi}{3} R^3}{A}\right)^{\frac{1}{3}} \approx 2 \text{ fm}, \tag{4.38}$$

a condition which leads to the relation

$$\frac{2\pi R}{2L} \geq d,$$

where L is the multipolarity of the surface mode. Making use of $R = 1.2 A^{1/3}$ one obtains

$$L \leq 2A^{\frac{1}{3}} \tag{4.39}$$

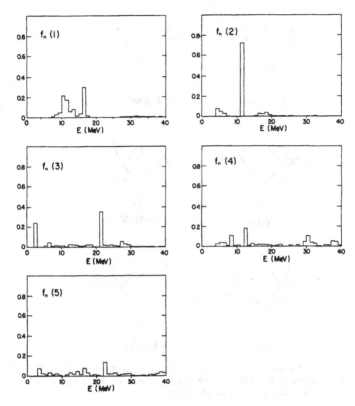

Figure 4.13. Energy and percentage of the energy weighted sum rule of the Random Phase Approximation solutions with multipolarities ranging from 1 to 5. The calculations were carried out making use of a Skyrme III interaction. (After Bortignon and Broglia (1981).)

which for ^{208}Pb leads to $L \leq 12$.

The residual interaction was diagonalized in the Random Phase Approximation in a basis of particle–hole excitations with energy ≤ 30 MeV. Results of this diagonalization are shown in Fig. 4.13. Of particular importance for the calculation of the damping width of giant resonances are the low-lying surface vibrational states: 3^- (2.6 MeV), 5^- (3.2 MeV) and 2^+ (4.07 MeV). All of them are rather collective, as testified by the corresponding percentage of the energy weighted sum rule they exhaust: 20%, 6% and 8% respectively. The giant quadrupole resonance at the level of RPA appears as a single peak at $\approx$ 12 MeV, exhausting $\approx$ 72% of the energy weighted sum rule. Making use of the above results, the self-energy of the giant quadrupole resonance was calculated. The particle-vibration coupling Hamiltonian defined in Eq. (3.20) was employed. For the average potential U, a Saxon–Woods parametrization of the Hartree–Fock potential was used.

In Fig. 4.14, typical amplitudes contributing to the giant quadrupole resonance self-energy are shown. The reason for considering in all cases the

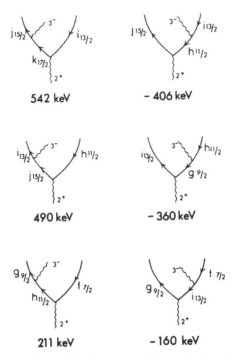

Figure 4.14.   Examples of cancellation between contributions to the matrix elements $V_{\alpha\alpha}$ associated with the coupling of the GQR of ^{208}Pb to 2p–2h states containing a low-lying collective surface vibration. (After Bortignon and Broglia (1981).)

low-lying octupole vibration as the α-boson (cf. Fig. 4.12) is because this is the single main coupling which contributes with about half of the total FWHM of the resonance. This is seen in Fig. 4.15 where the strength function (cf. Eq. (4.9)) of the giant quadrupole resonance calculated making use of the real and of the imaginary part of the self-energy are displayed. Also shown is the result including, as intermediate phonon in the calculation of the self-energy (α-phonon), only the low-lying 3^- surface vibration.

A strong cancellation, of the order of 50% or more, is found for the different intermediate states composed of an uncorrelated particle-hole excitation and of the octupole vibration (cf. Fig. 4.14). The quantitative relevance of this effect is better seen in Fig. 4.16, where the total squared matrix elements

$$V_{GQR,\alpha'}^2 = V_p^2 + V_h^2 + 2V_p V_h \times \text{(recoupling)}, \qquad (4.40)$$

with

$$V_p = X_{GR}(\nu_k, \nu_i) V(\nu_k, \nu_{k'}; \lambda)$$

and

$$V_h = X_{GR}(\nu_k, \nu_i) V(\nu_i, \nu_{i'}; \lambda)$$

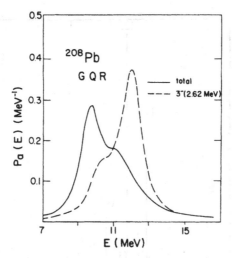

Figure 4.15. Strength function (cf. Eq. (4.9)) associated with the giant quadrupole resonance of ^{208}Pb taking into account couplings to all intermediate states containing surface vibrations (α-phonons, cf. Fig. 4.12) with multipolarity ≤ 10. Also shown are the results of a calculation where the only vibration allowed in the intermediate state is the low-lying 3^- state of ^{208}Pb. (After Bortignon and Broglia (1981).)

are displayed. The label $\alpha' \equiv \{(\nu_k \nu_{i'}, \lambda), (\nu_{k'} \nu_i, \lambda)\}$ refers to the quantum numbers of the intermediate "doorway state". The recoupling factor takes care of changing the coupling of the octupole phonon from the particle to the hole of the intermediate state and vice versa (cf. Figs. 4.12(c) and (d)).

From Fig. 4.16(b) one can estimate the average value of $V^2_{GQR,\alpha'}$ (≈ 0.02 MeV^2) and the density of 2p–2h intermediate states (≈ 20 MeV^{-1}). Using Fermi's Golden Rule, expressed in Eq. (4.5), one obtains

$$\Gamma^\downarrow_{GQR} \approx 2\pi V^2_{GQR,\alpha'} \varrho \approx 2.5 \text{ MeV}, \qquad (4.41)$$

in overall agreement with the value of the observed FWHM as well as with that obtained from the detailed calculation of the strength function.

We shall conclude the present section by remembering the solution of a puzzle regarding the properties of the giant quadrupole resonance of ^{208}Pb, which helps understanding the role doorway states play in the relaxation of giant resonances. It also sheds light into the interweaving of "doorway damping" and "compound nucleus damping". The two mechanisms constitute the first and the last step in the coupling of the giant resonance to progressively more complicate states (cf. Fig. 2.8 as well as Fig. 4.18). Otherwise, they are quite unrelated. Let us clarify this point with a Gedanken experiment, remembering that transitions leading to damping must be real, that is, they must conserve energy (on-the-energy-shell transitions).

Let us now assume a nucleus with an axially symmetric quadrupole deformation. According to the relation given by Eq. (3.53) (cf. also Sect. 10.6.1) this mode will split, at the level of mean field (RPA), into two peaks. One,

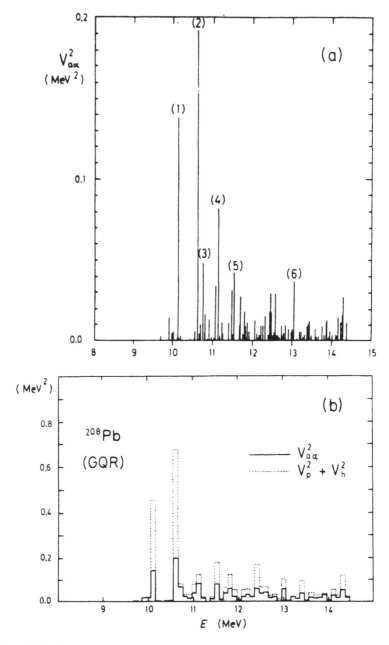

Figure 4.16. Spectral distribution of the square of the particle-vibration coupling matrix elements $V_{GQR,\alpha'}$ as a function of the energy of the intermediate doorway states α'. In (a) the doorway states $|1\rangle = |(3_1^-, h_{11/2}^{-1}(\pi))i_{13/2}(\pi)\rangle$, $|2\rangle = |(3_1^-, i_{11/2}^{-1}(\pi))j_{15/2}(\nu)\rangle$, $|3\rangle = |(3_1^-, d_{5/2}^{-1}(\pi))f_{7/2}(\pi)\rangle$, $|4\rangle = |(3_1^-, f_{7/2}^{-1}(\nu))g_{9/2}(\nu)\rangle$, $|5\rangle = |(5_1^-, i_{13/2}^{-1}(\nu))j_{15/2}(\nu)\rangle$, and $|6\rangle = |(7_1^-, i_{13/2}(\pi))j_{15/2}(\nu)\rangle$ are shown (cf. also Fig. 4.14). In (b) the same quantities are shown in histogram form (continuous line), in comparison with the sum $V_p^2 + V_h^2$ (cf. Eq. (4.40)) which is displayed with a dashed line. (After Bortignon and Broglia (1981).)

with an energy lower than $79A^{-\frac{1}{3}}$ MeV, corresponding to a vibration along the symmetry axis, and carrying $\frac{1}{3}$ of the oscillator strength. Another, at an energy higher than $79A^{-\frac{1}{3}}$ MeV, corresponding to vibrations in the plane perpendicular to the symmetry axis, and carrying $\frac{2}{3}$ of the oscillator strength. Let us furthermore assume that the deformation is so large that the lowest peak appears at a very low excitation energy, such that the only multiparticle–multihole configurations with energy similar to that of the lowest peak of the giant vibration, are states of the type $|(ph) \otimes \lambda\rangle$. That is, 2 particle–2 hole configurations consisting of an uncorrelated particle–hole excitation and of a surface vibration. In this case, only the first step of the hierarchy of couplings leading to the relaxation of giant vibrations is possible. The breaking of the strength will depend on the detailed structure of the single-particle motion and of surface vibrations, as well as on the particle-vibration coupling Hamiltonian. Consequently, the distribution of strength and intensity is expected to be "random". The situation is expected to be very different concerning the high energy peak. In this case a variety of many particle–many hole states will be present at similar excitation energy than that of the giant vibration, producing a dense background of states to which the vibration will couple. Even if each coupling may depend on detailed properties of the single-particle motion of the vibrations and of the coupling Hamiltonian, the complexity of the motion is expected to lead to a distribution of energy and strength which could also have been obtained by diagonalizing a random matrix. In keeping with this fact, the strength distribution is expected to follow a Porter–Thomas (cf., e.g., Bohr and Mottelson (1969)) distribution. It should be noted that the Gedanken experiment could become real if it became possible to measure the γ-decay of the giant dipole resonance based on a superdeformed or hyperdeformed configuration in rapidly rotating nuclei (M. Gallardo et al. (1985)).

Let us now return to the discussion of the giant quadrupole resonance in ^{208}Pb. In Fig. 4.17 we display the experimental data obtained in a high resolution electron scattering experiment. In the energy interval between 8 MeV and 12 MeV, about 60 peaks were observed. Theory predicts 60 peaks in this energy region, arising from "doorway coupling" and displaying a strength distribution which is similar to that observed experimentally (cf. Fig. 4.17). The scenario associated with the low energy peak of the resonance discussed in the Gedanken experiment comes to mind, and the above figure raised expectations on the possibility of a direct measurement of doorway splitting of giant resonances. The first indication that things did not work in this way came from the fact that, while theory predicted 60% of the EWSR in the energy interval under discussion, the analysis of the electron scattering data was consistent with only 30% of the EWSR. This figure was not only in disagreement with theory but, more important, with the results of hadron scattering experiments. These experiments confirmed the overall shape of the quadrupole strength distribution, but differed in a factor of 2 in the strength. The resolution of the puzzle was found (Kilgus et al. (1987)) by recognizing that the high density of levels

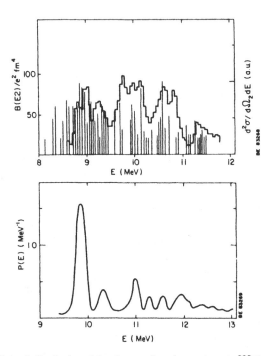

Figure 4.17. Strength distribution of the giant quadrupole resonance in ^{208}Pb. In the upper part, the (e,e′) data of Kühner et al. (1981) are displayed with a sharp line, together with the (α, α'n) data of Eyrich et al. (1982). In the lower part the theoretical strength function resulting from the coupling of the giant quadrupole resonance to doorway states (Bortignon and Broglia (1981)) is shown. Note the energy difference between the energy scale of the two parts of the figure. (From Bortignon et al. (1984).)

within the energy interval of interest, required a Porter–Thomas distribution of the strength. In other words, the scenario associated with the high-energy peak of the Gedanken experiment.

While in the case of giant resonances doorway damping and compound nucleus damping are likely to appear always together, we shall see that the two mechanisms could, in principle, be disentangled from each other, by measuring the damping of rotational motion (cf. Sect. 11.3.1).

We shall see in Sects. 4.3 and 4.4 that, in any case, it is still possible to gain insight into the "doorway" states to which the giant resonances couple, by looking at γ- and particle-decay of giant resonances (cf. also Sects. 2.4–2.6)

4.2.2 Compound coupling

In this section, we will study the "compound nucleus damping" mechanism, that is the contribution to the damping of the giant resonances due to the cou-

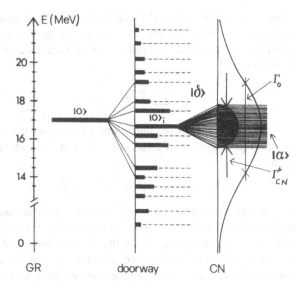

Figure 4.18. Schematic representation of the breaking of a giant resonance $|0\rangle$ due to coupling to doorway states $|\delta\rangle$, and decay in the compound nucleus (CN). The total width Γ_0 is essentially controlled by doorway coupling, while the compound damping ($\Gamma_{CN}^{\downarrow}$) is expected to provide a smoothing of the original distribution.

pling to progressively more complicated states up to the fully mixed compound nucleus states (cf. Lauritzen et al. (1995) and Fig. 4.18).

We write the total Hamiltonian describing the system as

$$H = H_0 + V, \qquad (4.42)$$

sum of a mean field Hamiltonian H_0, and a residual two-body interaction V. The eigenstates of the mean field Hamiltonian H_0 are the independent many-particle many-hole excitations of the system, that is the one-particle one-hole (1p–1h), 2p–2h, 3p–3h, etc. excitations. We shall start by diagonalizing the total Hamiltonian H in a restricted basis of 1p–1h states, obtaining the giant resonances (GR) plus low-lying collective states, aside from a large number of uncharacteristic states being close in energy to the unperturbed 1p–1h excitations.

The diagonalized 1p–1h states, together with the set of 2p–2h, 3p–3h etc. states form a complete set of basis states which we shall denote by $|\mu\rangle$, with energies $E_\mu = \langle \mu | H | \mu \rangle$. The exact stationary states of the total Hamiltonian obtained by diagonalizing H in the complete basis of states $|\mu\rangle$ will be denoted $|i\rangle$ with the energies E_i. The damping of the giant resonance state $|0\rangle$ is controlled by its coupling to the states $|\mu\rangle$. Among these states, in the 2p–2h class are the doorway states $|\delta_i\rangle$ considered in the previous section and consisting of a surface vibration plus an incoherent 1p–1h state. We shall show that the damping of the GR into the compound nucleus eigenstates can be well estimated just from its coupling to this set of doorway states.

For this purpose, we shall use a model based on the following assumptions. First, the spreading of any state $|\mu\rangle$ into the compound nucleus eigenstates $|i\rangle$ associated with the intrinsic chaotic dynamics is characterized by random, uncorrelated amplitudes X_μ^i,

$$|\mu\rangle = \sum_i X_\mu^i |i\rangle. \tag{4.43}$$

The second assumption is that of the uniformity of the spectrum and of the structure of the states $|\mu\rangle$. To each of these states we can associate a strength function written as

$$P_\mu(E_i) = \rho(E_i)|\langle\mu|i\rangle|^2 = \rho(E_i)|X_\mu^i|^2, \tag{4.44}$$

where ρ denotes the total density of states with the same quantum numbers as those of the mode under consideration. The assumption of uniformity asserts that $P_\mu(E_i) = P(E_i - E_\mu)$, where E_μ is the unperturbed energy of the state $|\mu\rangle$ and where it has been implicitly assumed an averaging over nearby states $|\mu\rangle$. In other words, the states $|\mu\rangle$ are all characterized by a single strength function P centered around each unperturbed energy E_μ. This in keeping with the fact that, for a single-particle damping width depending linearly on the energy (cf. Eq. (4.17)), the damping width of a many-particle-many-hole state is independent of the number of particle-hole excitations.

We cannot apply this argument in discussing the strength function $P_0(E - E_0)$ of the giant resonance state $|0\rangle$, because of the coherent properties of this mode, as already discussed in the previous section. As a result we allow the strength function P_0 of the giant resonance to be different from the strength function P.

The uniformity assumption also requires that the coupling between the states $|\mu\rangle$ have a uniform distribution. For the associated average coupling strength per unit energy, we write

$$\rho_\mu(E_\nu)\overline{|H_{\mu\nu}|^2} = \rho_\mu(E_\nu)\overline{|\langle\mu|H|\nu\rangle|^2} = f(E_\nu - E_\mu). \tag{4.45}$$

Here, ρ_μ denotes the level density of those states which couple, through the interaction V, to a given state $|\mu\rangle$. Similarly, for the GR we write,

$$\rho_0(E_\mu)\overline{|H_{0\mu}|^2} = \rho_0(E_\mu)\overline{|\langle0|H|\mu\rangle|^2} = f_0(E_\mu - E_0), \tag{4.46}$$

with ρ_0 being the density of doorway states associated with the state $|0\rangle$.

The diagonalization of the full Hamiltonian H now proceeds in two steps. First H is diagonalized among all the states $|\mu\rangle$ except for the GR state $|0\rangle$, thus producing a new set of eigenstates $|\alpha\rangle$. These states $|\alpha\rangle$ have no mutual interactions but couple only to the state $|0\rangle$ with typical matrix elements

$$\begin{aligned}
\overline{|H_{0\alpha}|^2} &= \overline{|\sum_\mu H_{0\mu}\langle\mu|\alpha\rangle|^2} \approx \sum_\mu \overline{|H_{0\mu}|^2}\ \overline{|\langle\mu|\alpha\rangle|^2} \\
&\approx \sum_\mu \overline{|H_{0\mu}|^2}\ \frac{1}{\rho(E_\alpha)}P_\mu(E_\alpha),
\end{aligned} \tag{4.47}$$

where the strength function P_μ is defined in Eq. (4.44). In writing the above equation, the following assumptions have been used: 1) that the amplitudes are uncorrelated, 2) that the omission of couplings between the states $|\mu\rangle$ and the GR state $|0\rangle$ does not affect the strength function $P_\mu(E)$ describing a typical state $|\mu\rangle$.

The coupling matrix elements $H_{0\alpha}$ control the last stages of the relaxation process leading to extremely complicated (chaotic) compound nucleus states $|i\rangle$, each carrying tiny fractions of the giant resonance strength. If N is the number of principal components in the compound nucleus states, then $\rho \sim N$ and, from Eq. (4.47), $H_{0\alpha} \sim N^{-1/2}$.

The second part of the diagonalization can be carried out exactly, with the strength function written as, cf. Bohr and Mottelson (1969),

$$P_0(E - E_0) = \frac{1}{\pi} \text{Im} \left[E - E_0 - \Sigma_0(E - E_0) \right]^{-1},$$

$$\Sigma_0(E - E_0) = \sum_\alpha \frac{|H_{0\alpha}|^2}{E - E_\alpha - i0}. \tag{4.48}$$

The self-energy part $\Sigma_0(E)$ of the giant resonance is expressed in terms of mixing matrix elements given in Eq. (4.47). Since we average over nearby shell model states we may write,

$$\begin{aligned}
\Sigma_0(E - E_0) &\approx \sum_\alpha \sum_\mu \frac{1}{E - E_\alpha - i0} \, \overline{|H_{0\mu}|^2} \frac{1}{\rho(E_\alpha)} P(E_\alpha - E_\mu) \\
&\approx \int dE_\alpha \int dE_\mu \left[\rho_0(E_\mu) \overline{|H_{0\mu}|^2} \right] \\
&\quad \times \frac{P(E_\alpha - E_\mu)}{E - E_\alpha - i0} \\
&= \int dx \int dy \, f_0(x - y) \frac{P(y)}{E - E_0 - x - i0}.
\end{aligned} \tag{4.49}$$

Equations (4.48) and (4.49) relate the strength function P_0 of the Giant resonance to the strength function P of the states $|\mu\rangle$.

One can use the same procedure of two-step diagonalization for any state $|\mu\rangle$. In keeping with Eq. (4.48), the strength function P_μ is written in terms of the corresponding self-energy Σ_μ which includes coupling to other states $|\nu\rangle$. Using the assumption of uniformity, the infinite set of equations for P_μ is substituted by the integral equation for the common strength function $P(E)$,

$$P(z) = \frac{1}{\pi} \text{Im} \left[z - \Sigma(z) \right]^{-1},$$

$$\Sigma(z) \approx \int dx \int dy \, f(x - y) \frac{P(y)}{z - x - i0}. \tag{4.50}$$

The set of coupled equations (4.48)–(4.50) can be solved by iteration and define the giant resonance strength function in terms of average coupling matrix elements introduced in Eq. (4.45) and (4.46). The relations (4.48)–(4.50) implicitly contain the coupling of the collective excitation to a hierarchy of ever more complicated many-particle many-hole states $|\mu\rangle$. In ordinary shell model calculations, the set of equations is truncated at a maximum number of particle–hole excitations. Here instead, we have imposed uniformity in the equations so that strength functions for high seniority states, i.e. large number of particle–hole excitations, are not neglected but rather retained at a limiting form.

The strength functions obtained by solving the set of coupled equations (4.48)–(4.50) depend both on the strength as well as the energy range of the residual interaction. In order to discuss some of the general properties of the solutions and eventually find analytic solutions to the equations, we parametrize the couplings given in Eq. (4.45) and (4.46) as

$$f_0(E) = \frac{n_0 v_0^2}{2\pi} \frac{W_0}{E^2 + (W_0/2)^2}, \tag{4.51}$$

and

$$f(E) = \frac{n v^2}{2\pi} \frac{W}{E^2 + (W/2)^2}. \tag{4.52}$$

The quantity W_0 measures the range in energies E_μ over which the giant resonance state $|0\rangle$ couples to the doorway states. The strength of the residual interaction $n_0 v_0^2$ has been written in terms of a typical r.m.s. matrix element v_0 and the number of doorway states n_0.

With this parametrization, the second moment of the strength function P_0 of the giant resonance $|0\rangle$ is given by

$$
\begin{aligned}
M_0^{(2)} &= \langle 0|(H - \langle H \rangle)^2|0\rangle = \sum_\mu |\langle 0|H|\mu\rangle|^2 \\
&\approx \int dE_\mu \, f_0(E_\mu - E_0) = \sum_{\mu=1}^{n_0} v_0^2 = n_0 v_0^2, \tag{4.53}
\end{aligned}
$$

and a similar expression holds for the second moment $M_\mu^{(2)} (= \sum_{\mu=1}^{n} v^2)$, of any state $|\mu\rangle$.

With the above parametrization, the self-energies acquire the form

$$
\begin{aligned}
\Sigma_0(z) &= \int dx \, \frac{n_0 v_0^2}{z - x - iW_0/2} \, P(x), \\
\Sigma(z) &= \int dx \, \frac{n v^2}{z - x - iW/2} \, P(x). \tag{4.54}
\end{aligned}
$$

In the weak coupling limit, $\sqrt{n_0} v_0 \ll W_0$ and $\sqrt{n} v \ll W$, the self energy $\Sigma_0(z)$ is a slowly changing function of energy within the spreading width. Consequently, the collective strength function will not depend on the details of $P(x)$ and eventually attain a Breit–Wigner form (exponential decay),

$$P_0(E - E_0) = \frac{1}{2\pi} \frac{\Gamma_0}{(E - E_0)^2 + (\Gamma_0/2)^2}, \tag{4.55}$$

where the spreading width of the giant resonance is given by

$$\Gamma_0 = 2 \, \mathrm{Im}\Sigma_0(0) = \frac{4n_0 v_0^2}{W_0}; \qquad \sqrt{n_0} v_0 \ll W_0. \qquad (4.56)$$

This result also could have been obtained directly from Eq. (4.46) and (4.51) using Fermi's Golden Rule, that is,

$$\Gamma_0 = 2\pi v_0^2 \rho_0(E_\mu \approx E_0) = 2\pi f_0(0). \qquad (4.57)$$

In keeping with the assumptions made above, this relation applies in those cases in which the damping width Γ_0 is much smaller than the range of the interaction W_0.

In the strong coupling limit for the intrinsic states, that is when $\sqrt{n}v \gg W$, the exact shape of the coupling parametrized in Eqs. (4.51) and (4.52) is not important and the Hamiltonian can be modelled by a banded matrix with random matrix elements of magnitude v in a band of size n surrounding the diagonal. The resulting self-consistent strength function of generic background states is a semicircle of radius R, that is

$$P(z) = \frac{2}{\pi R^2}\sqrt{R^2 - z^2}, \quad R = 2\sqrt{n}v; \quad \sqrt{n}v \gg W. \qquad (4.58)$$

The giant resonance strength function, solution of Eq. (4.54) and (4.58) becomes

$$P_0(z) = \frac{2}{\pi} \frac{\sqrt{R^2 - z^2}}{R_0^2 + 4z^2(R^2/R_0^2 - 1)}, \qquad (4.59)$$

where $R_0 = 2\sqrt{n_0}v_0$, that is, a hybrid of the semicircle and Breit–Wigner shapes. The effective spreading width (FWHM) of the giant resonance varies from $\Gamma_0 = \sqrt{3}R_0$ at $R_0 = R$ to $\Gamma_0 = R_0^2/R$ at $R \gg R_0$. Again, the upper boundary for the width is determined solely by the coupling to the doorway states. Subsequent scattering into the compound nucleus eigenstates may change the shape of the strength function but does not add to the damping width which may actually decrease. This effect is a reminder of motional narrowing, cf. Sect. 10.7 and Chapter 11. Connected with this phenomenon there is a change in the structure of the strength function, from one with its wings stretched out (Breit–Wigner) to one that is confined (Wigner). Note that the result given by Eq. (4.59) (strong coupling limit) could not have been obtained in perturbation theory.

The estimates of P_0 collected in Eqs. (4.55) and (4.59), give an upper limit for the damping width of the giant resonances. These results are expected to remain qualitatively valid also in the case where the compound nucleus has a very high excitation energy and the giant resonance is thermally excited, provided a thermal averaging of Eq. (4.48)–(4.50) is carried out, cf. Chs. 9 and 10. This is because, the quantity $R_0 = 2\sqrt{n_0}v_0$, which is proportional to the square root of the expectation value of H^2 in the giant resonance state, cf. Eq. (4.53), does not contain any exponentially growing parameters. On the other hand, both n_0 and v_0 change with temperature, as discussed in Sect. 8.4.2.

Making use of the example of the giant quadrupole resonance of ^{208}Pb, one obtains from Eq. (4.57) ($n_0 \approx 60$, $v^2 \approx 0.02$ MeV^2, $W_0 \approx 2$ MeV; cf. Fig. 4.16(b)) $\Gamma_0 \approx 2.4$ MeV. This number is essentially equal to $\Gamma^{\downarrow}_{GQR}$ (cf. Eq. (4.41). This is because to couple to the compound nucleus states, the giant resonance has first to couple to the doorway states. Furthermore, because the coupling matrix elements $H_{0\alpha}$ to the compound state scale with $1/\sqrt{N}$, and thus neutralize the high density of compound nuclear levels in the corresponding expression of the damping width. The coupling to the compound nucleus thus provides a smooth averaging of the doorway states ($\Gamma^{\downarrow}_{CN} \approx 2.4$ MeV/60≈ 40 keV) and a physical interpretation of the parameter I introduced after Eq. (4.1). In other words, both the centroid of the giant resonances as well as the FWHM are determined by the detailed dynamics of the problem, namely shell structure and doorway coupling. On the other hand the fluctuations in the energies and in intensities are expected to display Wigner and Porter–Thomas behaviour, respectively.

This was, in synthesis, the basis for the reanalysis of the electron scattering experiments carried out by Kilgus et al. (1987), which removed the discrepancy between the electron and the hadron data as mentioned at the end of subsection 4.2.1. In fact, assuming a Porter–Thomas distribution for the strength of the different lines, part of what had been considered before background became giant quadrupole resonance. At this point, the resulting cross sections agreed, within the experimental errors, with the one obtained from hadron scattering.

4.3 Particle-Decay of Giant Resonances

In this section, we present the theoretical basis for the analysis of the experimental data associated with the neutron decay of ^{208}Pb discussed in Sect. 2.5. Use is made of the projection operator reaction theory of Feshbach et al. (1967). In this approach, the continuum–RPA spectrum is described in terms of discrete complex states. The imaginary part of their complex energies is related to their escape widths.

4.3.1 The continuum–RPA approach

We start by briefly deriving the RPA equations in coordinate space, following Bertsch and Tsai (1975).

Given the Hartree–Fock Hamiltonian H_0, which is a functional of the density ρ, we add a time-dependent (harmonic) perturbation

$$V_{ext}(\vec{r}, t) = V_{ext} \exp\left(-i\omega t\right) + H.C., \tag{4.60}$$

The perturbation produces a change of the density

$$\delta\rho(\vec{r}, t) = \delta\rho(\vec{r}) \exp(-i\omega t) + H.C. \qquad (4.61)$$

and the total Hamiltonian is

$$H(t) = H_0 + [(V_{ext} + \frac{\delta H_0}{\delta\rho}\delta\rho)\exp(-i\omega t) + H.C.]. \qquad (4.62)$$

The Hartree–Fock wavefunctions $\psi_i^{(0)}$ are also modified according to

$$\psi_i(\vec{r}, t) = \exp(-i\epsilon_i t)(\psi_i^{(0)}(\vec{r}) + \delta\psi_i^{(-)}(\vec{r})\exp(-i\omega t)$$
$$+ \delta\psi_i^{(+)}(\vec{r})\exp(+i\omega t)). \qquad (4.63)$$

The function $\delta\rho(\vec{r})$ in Eq. (4.61) is given, to first order, by

$$\delta\rho(\vec{r}) = \sum_i \psi_i^{(0)*}\delta\psi_i^{(-)} + \delta\psi_i^{(+)}\psi_i^{(0)*}. \qquad (4.64)$$

A formal solution of the Schrödinger equation for the functions $\delta\psi_i^{(\pm)}$ may be written as

$$\delta\psi_i^{(-)} = \frac{-1}{H_0 - \epsilon_i - \hbar\omega}(V_{ext} + \frac{\delta H}{\delta\rho}\delta\rho)\psi_i^{(0)}, \qquad (4.65)$$

and

$$\delta\psi_i^{(+)} = \frac{-1}{H_0 - \epsilon_i + \hbar\omega}(V_{ext} + \frac{\delta H}{\delta\rho}\delta\rho)^\dagger\psi_i^{(0)}. \qquad (4.66)$$

Substituting these solutions in the equation for $\delta\rho$, we may write compactly

$$\delta\rho = -V_{ext}G_{RPA}, \qquad (4.67)$$

where

$$G_{RPA}(\vec{r}_1, \vec{r}_2; \omega) = (1 + Gph\frac{\delta H}{\delta\rho})^{-1}G_{ph}, \qquad (4.68)$$

having defined the unperturbed particle–hole Green's function,

$$G_{ph}(\vec{r}_1, \vec{r}_2; \omega) = \sum_i \psi_i^{(0)*}(\vec{r}_1)[\frac{1}{H_0 - \epsilon_i - \hbar\omega}$$
$$+ \frac{1}{H_0 - \epsilon_i + \hbar\omega}]\psi_i^{(0)}(\vec{r}_2). \qquad (4.69)$$

The continuum–RPA approach consists now in solving by matrix inversion, in coordinate–space representation, Eq. (4.68) for G_{RPA}. After a multipole decomposition, its radial part is evaluated by the usual Green's function formula for second-order linear differential equations,

$$[H_0 - \epsilon]_{r_1, r_2} = \frac{u(r_<)w(r_>)}{W}. \qquad (4.70)$$

Here u and w are solutions of the radial differential equations with appropriate boundary conditions. The wave function u must be regular at the origin and

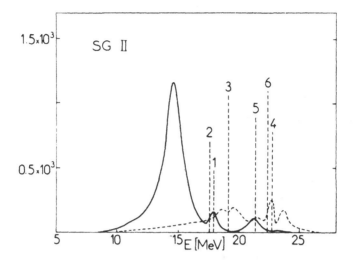

Figure 4.19. Monopole strength distribution in ^{208}Pb calculated in Nguyen van Giai and Sagawa (1981) with the SGII interaction. The RPA (solid line) and the unperturbed (dashed line) strengths are displayed, while vertical lines indicate the positions of discrete, unperturbed p–h configurations.

w must be a pure outgoing wave at infinity, while W is the Wronskian of the differential operator. The excited states of the system correspond to the poles of G_{RPA}, and for the strength function (cf. also Eq. (4.9))

$$S(\omega) = \sum_f \langle f|V_{ext}|0\rangle^2 \delta(E_f - E_0 - \omega), \qquad (4.71)$$

we may write

$$S(\omega) = \int d\vec{r}_1 d\vec{r}_2 V_{ext}(\vec{r}_1) Im G_{RPA}(\vec{r}_1, \vec{r}_2; \omega) V_{ext}(\vec{r}_2). \qquad (4.72)$$

As an example, the strength function for the isoscalar Giant Monopole Resonance in ^{208}Pb is shown in Fig. 4.19.

4.3.2 The doorway-state projection-operator method

The basic ideas of the projection operator method of Feshbach et al. (1967) are easily discussed in the single-particle case, with the help of the Fig. 4.20, from Nguyen van Giai et al. (1988).

Let us consider a particle moving in a finite, real potential well $U(r)$. The spectrum of the full Hamiltonian H of the system includes a number of bound states and a continuum of states above threshold. This continuum may display single-particle resonances, quasi-bound states, due to centrifugal and Coulomb effects (cf. Fig. 4.20A). The corresponding strength function $S(E)$ for an operator creating particle–hole excitations is sketched in Fig. 4.20B. Discrete

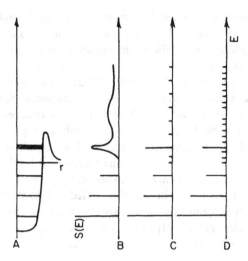

Figure 4.20. A) Spectrum of a finite potential well; B) exact strength distribution for a particle–hole operator; C) discrete strength distribution calculated in the space Q1; D) the same in a space Q2 larger than Q1. (Adapted from Nguyen van Giai et al. (1988).)

spikes, resonance peaks with a width of the order of the single-particle width, and a smooth distribution at higher excitation energy are observed. According to the projection operator method, a discrete subspace $Q1$ of finite dimension n_1 is defined, within which the hamiltonian H is diagonalized. For a sensible choice of Q_1, all the exact bound and quasi-bound states of H may be correctly approximated by the eigenstates of $Q_1 H Q_1$. Besides these physical states, there will be a background of other states above threshold without physical meaning, but necessary to make a complete basis in Q_1, as shown in Fig. 4.20C. In a subspace Q_2 larger than Q_1, the positions and level density of the background states will change, but the physical states will remain unaffected, as shown in Fig. 4.20D. This is obvious for the bound states, and is also valid for the quasi-bound states, since they have wave functions concentrated in the nuclear interior.

The construction of the unbound states of the subspace P, complementary of Q_1, runs as follows (cf., e.g., Colò et al. (1994) and refs. therein). At positive energy ϵ the radial scattering equation for H is solved for each partial wave $c = (l, j)$ projected on the orthogonal complement of Q_1, that is,

$$(H_{PP} - \varepsilon)u_{c,\varepsilon} = 0, \tag{4.73}$$

where

$$H_{PP} = (1 - \sum_i |\varphi_i\rangle\langle|\varphi_i|)H(1 - \sum_j |\varphi_j\rangle\langle|\varphi_j|), \tag{4.74}$$

with the sums extending over all the states of the subspace Q_1. In this way, it is ensured that the outgoing wave functions $u_{c,\varepsilon}^{(+)}$ have no overlap with any states of Q_1 and have no resonant behaviour, as all quasi-bound states belong to Q_1. The coupling through H of the states belonging to the subspace Q_1 with the scattering states of P, allows the infinitely long-lived states of Q_1 to decay by particle emission.

In most of the applications (Colò et al. (1992) and (1994)), Hartree–Fock equations have been solved using a two-body interaction of Skyrme type and the resulting self-consistent mean field have been diagonalized on a basis of harmonic oscillator wave functions. This diagonalization procedure can also be replaced by a diagonalization of the Hartree–Fock equations within a box. The result is a discrete set of occupied and unoccupied levels, including the quasi-bound levels at positive energy. An appropriate cutoff to describe each concrete situation, leads to a finite set of occupied and unoccupied levels $|i\rangle$.

The subspace Q_1 is formed by the Hartree–Fock ground state and by all the possible one particle–one hole (1p–1h) excitations within this set of Hartree–Fock states. The nuclear Hamiltonian is written as

$$H = H_0 + V_{ph}, \tag{4.75}$$

where H_0 is the Hartree–Fock Hamiltonian and V_{ph} is the particle–hole interaction determined as the functional derivative of the self-consistent mean field with respect to the density. The random phase approximation in a discrete particle–hole space amounts now to solve the corresponding equations with the projected Hamiltonian $Q_1 H Q_1$.

In order to account for escape and spreading effects, two other subspaces P and Q_2 are built. The subspace P contains the particle–hole configurations with the particle in an unbound state, orthogonal to all states $|i\rangle$. To determine these unbound states, the procedure described in connection with Eq. (4.73) is used. The space Q_2 is built with a set of "doorway states" $|D\rangle$, containing a 1p–1h excitation coupled to a collective vibration as in Sect. 4.2.

The nuclear Green's function G for a nucleus excited at an energy ω, satisfies the equation

$$(\omega - H + i\epsilon)G = 1. \tag{4.76}$$

It may be written as a sum of terms like $Q_1 G Q_1 + Q_1 G P + \cdots$, where the Green's function $Q_1 G Q_1$ can be shown to obey an equation where the effective Hamiltonian, after truncation of higher order couplings, is (Yoshida (1983)),

$$\begin{aligned}
\mathcal{H}(\omega) &\equiv Q_1 H Q_1 + W^\uparrow(\omega) + W^\downarrow(\omega) \\
&= Q_1 H Q_1 + Q_1 H P \frac{1}{\omega - PHP + i\epsilon} P H Q_1 \\
&+ Q_1 H Q_2 \frac{1}{\omega - Q_2 H Q_2 + i\epsilon} Q_2 H Q_1.
\end{aligned} \tag{4.77}$$

This energy-dependent, complex Hamiltonian allows to work inside the space Q_1. It has complex eigenvalues, the corresponding imaginary parts originating from the coupling of the space Q_1 to unbound and to more complicated configurations give rise to escape and spreading widths.

The escape term $W^\uparrow(\omega)$ can be evaluated in rather simple terms, by replacing the total Hamiltonian H by the one-body Hamiltonian H_0. The neglect of the matrix elements of $Q_1 V_{ph} P$ is expected to be rather safe, since discrete and continuum wave functions are essentially restricted to different radial intervals while the interaction V_{ph} has zero range. In this approximation, the escape term can be written (Yoshida and Adachi (1986)) as,

$$W^\uparrow(\omega) = \omega - H_0 - K, \qquad (4.78)$$

where K is the inverse, inside subspace Q_1, of the Green's function containing only the mean field Hamiltonian. This Green's function can be computed and inverted with ease, because it has only one-body matrix elements. The accuracy of the procedure has been checked in Nguyen van Giai et al. (1987), by comparing the isoscalar monopole strength distribution in ^{40}Ca, calculated using the approximate escape term $W^\uparrow$, with that obtained from exact continuum-RPA calculations. Good agreement between the two calculations was found (cf., e.g., Fig. 1 of Nguyen van Giai et al. (1987)). Similar comparisons have been made by Yoshida and Adachi (1986a), with similar conclusions.

The matrix elements of the spreading term $W^\downarrow(\omega)$ in the basis of states belonging to the subspace $Q1$ are given by

$$W^\downarrow_{ph,p'h'}(\omega) = \sum_D \frac{\langle ph|V_{ph}|D\rangle\langle D|V_{ph}|p'h'\rangle}{\omega - \omega_D}, \qquad (4.79)$$

where ω_D is the energy of the state $|D\rangle$, calculated in mean field approximation.

To solve the effective Hamiltonian (4.77), it is convenient to work on the basis of the RPA states, labelled by $|n\rangle$, obtained by diagonalizing its first term. The eigenvalue equation for the effective Hamiltonian defined in Eq. (4.77) can then be written as

$$\begin{pmatrix} \mathcal{D} + \mathcal{A}_1(\omega) & \mathcal{A}_2(\omega) \\ \mathcal{A}_3(\omega) & -\mathcal{D} + \mathcal{A}_4(\omega) \end{pmatrix} \begin{pmatrix} F^{(\nu)} \\ \bar{F}^{(\nu)} \end{pmatrix} = \left(\Omega_\nu - i\frac{\Gamma_\nu}{2}\right) \begin{pmatrix} F^{(\nu)} \\ \bar{F}^{(\nu)} \end{pmatrix},$$
$$(4.80)$$

where $\mathcal{D}$ is a diagonal matrix with the RPA eigenvalues, and the energy dependent matrices $\mathcal{A}_i$ contain the escape and spreading contributions. The energy dependence of eigenvalues and eigenvectors in Eq. (4.80) have been omitted for simplicity, while the amplitudes corresponding to positive and negative RPA eigenstates are collectively denoted F and $\bar{F}$ respectively. The solutions of Eq. (4.80) can be written as

$$|\nu\rangle = \sum_n F_n^{(\nu)}|n\rangle. \qquad (4.81)$$

The matrix which appears in (4.80) is complex symmetric. The transformation which makes it diagonal, that is the matrix of its eigenvectors, is complex orthogonal, that is

$$F^T F = F F^T = 1. \tag{4.82}$$

In terms of solutions of Eq. (4.80), the strength function is

$$S(\omega) = -\frac{1}{\pi} Im \sum_{\nu} \langle 0|O|\nu \rangle^2 \frac{1}{\omega - \Omega_\nu + i\frac{\Gamma_\nu}{2}}, \tag{4.83}$$

where the squared matrix element of O appears, instead of its squared modulus, due to the properties of the eigenvectors $|\nu\rangle$, which form a biorthogonal basis.

Another quantity which can be extracted from the model and which is actually measured in the particle decay experiments, is the branching ratio corresponding to a particular decay channel and defined as

$$B_c(\omega) \equiv \frac{\sigma_c(\omega)}{\sigma_{exc}(\omega)},$$

$$= \frac{\sum_{\nu,\nu'} S_{\nu'\nu} \gamma_{\nu'\nu,c} (\omega - \Omega_\nu - i\frac{\Gamma_\nu}{2})^{-1} (\omega - \Omega_{\nu'} + i\frac{\Gamma_{\nu'}}{2})^{-1}}{-2Im \sum_{\nu,\nu'} S_{\nu'\nu} (F^* F^T)_{\nu\nu'} (\omega - \Omega_{\nu'} - i\frac{\Gamma_{\nu'}}{2})^{-1}} \tag{4.84}$$

where $S_{\nu\nu'}$ is given by

$$S_{\nu\nu'} = \langle \nu|O|0 \rangle \langle \nu'|O|0 \rangle^*. \tag{4.85}$$

The energy $\omega = \varepsilon - \varepsilon_h$ is the difference between the energy ε of the escaping nucleon from the system A and the energy ε_h of the hole state in the residual $(A - 1)$ nucleus. The quantity $\gamma_{\nu\nu',c}$ is defined as

$$\gamma_{\nu\nu',c} = \int d\Omega_k \gamma_{\nu,c}(\vec{k}) \gamma_{\nu',c}^*(\vec{k}), \tag{4.86}$$

with

$$\gamma_{\nu,c}(\vec{k}) = \langle \varphi_c \, u_{c,\varepsilon}^{(-)}(\vec{k})|H_0|\nu \rangle, \tag{4.87}$$

where φ_c is the wave function describing the residual $(A - 1)$ nucleus in channel c, and $u_{c,\varepsilon}^{(-)}(\vec{k})$ is the escaping particle wavefunction belonging to space P.

Results of the calculation of the neutron decay of the giant monopole resonance of ^{208}Pb, carried out within the scheme discussed above, are compared in Fig. 4.21 with the experimental data (cf. Table 2.1).

The values of the centroid energy and of the width (FWHM) of the calculated strength function are in good agreement with the experimental data (13.9 MeV and 2.9 MeV respectively). The coupling to the doorway states changes the mean energy by about 1 MeV, corresponding to a reduction of K_A (cf. Sect. 2.3) of the order of 15%, much smaller than the effect advocated in Brown and Osnes (1985) to bring down K_{nm} to a value of ≈ 100 MeV.

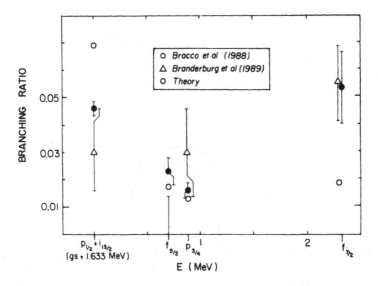

Figure 4.21. Branching ratios 'of neutron decay from the giant monopole resonance of ^{208}Pb. The experimental values are from Bracco et al. (1988) (solid dots) and from Brandenburg et al. (1989) (open triangle). The theoretical results are from Colò et al. (1992) (open dots).

4.4 Gamma-Decay of Giant Resonances

The gamma-decay of giant resonances to low-lying states can provide information concerning the coupling of giant modes to low-frequency collective surface vibrations (cf. Sect. 2.6). This possibility is illustrated in Fig. 4.22, where the processes associated with the E1-decay of a giant quadrupole resonance (GQR) to a low-lying octupole surface vibration (3^- state) are displayed. In (A), the γ-ray interacts directly with a particle-hole vertex or it can first excite the giant dipole resonance (GDR) which decays into a particle–hole pair. Thus the second graph in (A), with the GDR phonon drawn as a horizontal line, stands for all time orderings of the action of the external field. Such coupling leads to the effective charge (Bohr and Mottelson (1975)),

$$e_{eff} = e[1 + \chi(\Delta E)], \tag{4.88}$$

where the dipole polarizability is given by

$$\chi(\Delta E) = -0.76(\hbar\omega_D) \times$$

$$\times [(\hbar\omega_D + \Delta E + i\Gamma_D/2)(\hbar\omega_D - \Delta E - i\Gamma_D/2)]^{-1}. \tag{4.89}$$

In the above equation, ΔE is the energy of the dipole transition from the GQR to the low-lying octupole state. A similar discussion applies to the processes described in (B). Interchanging the role of particles and holes, graphs (C) and (D) are obtained. Clearly, the diagrams in (A) and in (C) have a very similar

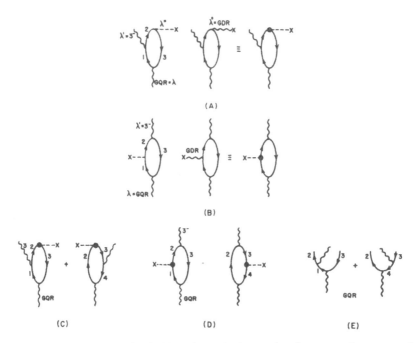

Figure 4.22. Processes associated with the decay of a giant quadrupole resonance into an octupole vibration. See text. (Adapted from Bortignon et al. (1984a).)

structure to that of the graphs entering in the calculation of the strength function of giant resonances, as shown in (E) (cf. Figs. 4.12 and 4.14).

Once the matrix element of the dipole operator $M_{f,i}$ is calculated taking into account all the contributions displayed in Fig. 4.22, the gamma-width for the transition can be written as

$$d\Gamma_\gamma(GQR(E) \to 3^-)/dE \approx$$

$$\approx |1 + \chi(\Delta E)|^2 (\Delta E)^3 |M_{3^-,GQR}|^2 P_{GQR}(E), \qquad (4.90)$$

where $P_{GQR}(E)$ is the GQR strength function (cf. Fig. 4.15) for the nucleus ^{208}Pb. Making use of the results displayed with the continuous curve, one obtains for the width Γ_γ integrated over the GQR energy region, a value of 3.5 eV, in overall agreement with the experimental finding of Bertrand et al. (1984). This value is a factor 500 smaller than a typical particle–hole dipole transition of the same transition energy (≈ 8 MeV). This strong reduction of the decay width, is due to the isovector nature of the dipole operator connecting two states of predominantly isoscalar character (proton and neutron contributions tend to cancel) and to the strong screening factor, $|1 + \chi|^2 \approx 0.07$, produced by the excitation of the GDR for a transition of 8 MeV energy. A further reduction factor is found in the destructive interference between particle- and hole-contributions to the process (that is, between the contributions associated with graphs (C) and (D)).

The first two reduction factors are not operative for the E1-transition connecting the isovector giant quadrupole resonance (at about 20 MeV in ^{208}Pb) and the low-lying octupole state. Profiting from this fact, the corresponding gamma-decay was used (Beene et al. (1988)) to study the strength distribution of this resonance (cf. also Speth et al. (1985)).

Part 2

Finite Temperature

5

Measurement of Giant Resonances

The question of how nuclear motion is affected by the internal excitation energy of the system is the main theme of the present monograph. In this chapter we shall introduce the subject by reviewing the experimental situation regarding the giant dipole resonance, the single giant vibration which has been clearly identified in the γ-decay of excited nuclei.

As a rule, the density of levels $\rho(A, E^*)$ of a nucleus with A nucleons, at an excitation energy E^* is very high as compared to the energy resolution with which one can specify the state of the system. In fact, using Bethe's Fermi-gas formula[1]

$$\rho(A, E^*) \approx \frac{e^{2\sqrt{aE^*}}}{\sqrt{48}E^*}, \tag{5.1}$$

one obtains 10^{17} states per MeV, for a nucleus of mass number A = 100 at an excitation energy of 50 MeV. In this estimate we have used the empirical value (cf. Sect. 7.2.1)

$$a \approx A/10 \text{ MeV}^{-1}, \tag{5.2}$$

of the level density parameter consistent with the observations carried out at low and moderate excitation energies ($E^* \leq 100$ MeV). This result makes the concept of temperature

$$T = \left(\frac{1}{\rho}\frac{d\rho}{dE^*}\right)^{-1}, \tag{5.3}$$

well defined, also in the case of a system, like the nucleus, with a finite number of particles (cf. also Sect. 1.2 and Ch. 7). The fact that the level density displays an exponential dependence with the excitation energy implies that also the energy spectra of γ-rays arising from the decay of a compound

---

[1] There are a number of expressions of the type given in Eq. (5.1), differing only in the prefactor (cf., e.g., Eq. (5.11) and also Ch. 7).

nucleus will have an exponential form. In fact, as it will be discussed in Sect. 5.4, in an equilibrated system at temperature T, the probability for the production of particles (photons) of energy E_γ follows the law $\exp(-E_\gamma/T)$.

5.1 Overview

The first evidence for the existence of the giant dipole oscillations in excited nuclei was found in the early 1980s in Berkeley in measurements of spectra of γ-rays with transition energies between 5 to 25 MeV emitted by compound nuclei following fusion reactions induced in heavy ion collisions (Newton et al. (1981)). The giant dipole resonance was seen as a change in the slope of the intensity curve taking place in the range of the resonance energy (cf. Fig. 5.1(a)). In Fig. 5.1(b) the variety of decays which cools the compound nucleus are shown schematically. Giant-dipole-photons indicated by the label (1) in the figure are emitted in an early stage in competition with neutrons. When the intrinsic energy of the system becomes lower than the neutron separation energy, slower γ-transitions have time to occur, that is, transitions of type (2) in Fig. 5.1(b). Most of the angular momentum is carried off in the final stage of the decay by quadrupole radiation, indicated by (3) in Fig. 5.1(b). The different probabilities associated with the different decays are discussed in Sect. 5.4. In any case, because the dipole energy is of the order of 15 MeV, and the temperature of the compound nucleus at which giant resonances have been observed lies in the energy range 1–3 MeV, the probability of thermally exciting a dipole state is 10^{-5}–10^{-3}.

Another evidence for the presence of giant dipole vibrations built on excited states was obtained only a few years after of the Berkeley experiment in measurements of proton capture cross section. Studies of reactions of the type $A - 1(p, \gamma_i)A^*$ as a function of proton bombarding energy, and leading to a given final excited state in the residual nucleus A, have shown resonant behaviour at excitation energies equal to the sum of the energy of the final state plus the energy of the giant dipole resonance (Snover (1986)).

Since the time in which the presence of the giant dipole resonance in hot nuclei as a thermal excitation was established, much experimental and theoretical work has been dedicated to the subject (cf., e.g., the Proceedings of the Topical Conferences on giant resonances which have taken place in Legnaro (1987), Gull Lake (1993), Groningen (1995)). The experimental findings are mainly based on measurements of γ-decay following fusion reactions induced by heavy-ion collisions. The basic information from these studies are: the photoemission cross section integrated over the energy (oscillator strength), the centroid energy of the resonance and the resonance width. A compilation of these results is reported in Figs. 5.2(a)–(e) as a function of the nucleus mass number. Recently, the angular distribution of the photoemission cross section has also been measured (cf. Sect. 6.1.2)).

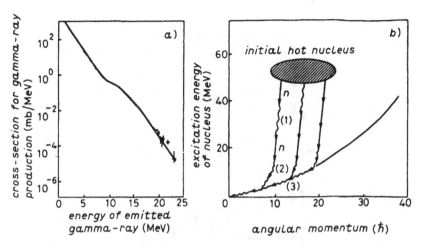

Figure 5.1. (a) A typical measured γ-ray spectrum from the decay of a compound nucleus formed in a heavy-ion-induced fusion reaction (adapted from Gosset et al. (1985)). The decay is shown schematically in (b). Giant-dipole-oscillation photons, indicated by (1), are emitted in an early stage in competition with neutrons. Following neutron emission, slower γ-transitions have time to occur (2). Most of the angular momentum is carried off in the final decay by quadrupole radiation, indicated in (3).

It is interesting to make a comparison between Figs. 5.2(a)–(e) and the corresponding Figs. 2.5(a)–(e) displaying the properties of the giant dipole resonance based on the ground state. A summary of such a comparison, which averages over mass number as well as over excitation energy and rotational frequency, is displayed in Fig. 5.3. The main feature emerging from this comparison is that the energy weighted sum rule and the energy centroid essentially do not depend on temperature. In keeping with the fact that in most cases the compound nucleus is strongly rotating, these quantities do not seem to depend on angular momentum. On the other hand, conspicuous differences are observed in the total width of the resonance (cf. Figs. 2.5(c) and 5.2(c)), in the dispersion of the mean value of the effective quadrupole deformation parameter (cf. Figs. 2.5(d) and 5.2(d)) and in the ratio of the upper to lower components of the two peak resonances (cf. Figs. 2.5(e) and 5.2(e)). In fact, while the average value of the total width of the giant dipole resonance based on the ground state is ≈ 5.8 MeV, that associated with thermally excited dipole vibrations in hot, rapidly rotating nuclei is ≈ 10 MeV. Furthermore, the strong shell structure displayed by the total width of the giant dipole resonance in cold nuclei is absent from the corresponding quantity measured in the γ-decay of compound nuclei. Concerning the dispersion of the mean value of the effective quadrupole deformation parameter, one observes a value of 0.2 in cold nuclei and of 0.36 in compound nuclei. Again, the strong shell structure displayed by the values associated with cold nuclei is absent from the compound nucleus data. Of the 28 experimental values of the ratio of strengths associated with the upper and the lower Lorentzian components used to fit the giant dipole

EXCITED STATE GDR

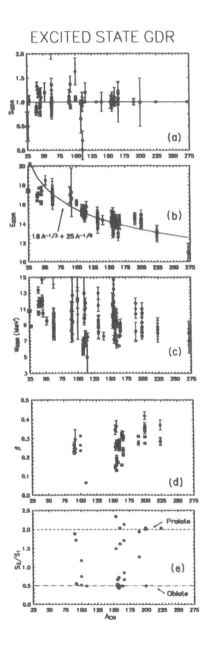

Figure 5.2. Observed properties describing the giant dipole resonance based on excited states as a function of the nuclear mass number A: (a) strength in units of the energy weighted sum rule, (b) centroid energy, (c) FWHM, (d) quadrupole deformation parameter β as obtained from the giant dipole resonance lineshape (cf. Eq. (2.2)), (e) ratio of the strength S_2 associated with the high-energy component and of the strength S_1 of the low-energy component. (Adapted from Gaardhøje (1992).)

strength function based on the ground state, only 6 (20%) fall within 20% of the limiting values corresponding to oblate (0.5 ± 0.1) and prolate (2 ± 0.4) nuclear shapes in the case of cold nuclei (cf. Fig. 2.5(e)). For hot, strongly rotating nuclei out of the 28 cases, 18 ($\approx 60\%$) fall within these limiting values (cf. Fig. 5.2(e)).

It is not likely that with the present knowledge on nuclear structure, a detailed account of each experimental point can be obtained. On the other hand, as we shall see, theory seems to be able to provide an overall account of the main trends of the experimental findings.

The independence of the oscillator strength and of the energy centroid (cf. Figs. 5.3(a) and (b)) with temperature and angular momentum can be understood in terms of the resilience of the mean field, where nucleons move essentially independently from each other, with respect to those quantities (cf. Ch. 8). In fact, the parameter which controls the rigidity of the mean field is $\hbar\omega_0$ ($\approx 41\, A^{-1/3}$ MeV), that is the energy difference between major shells. This quantity is 9–7 MeV for medium-heavy nuclei ($100 \leq A \leq 200$), an energy which has to be compared with values of the temperature of the order 2–3 MeV and rotational frequencies $\hbar\omega_{rot} \approx 1$ MeV. The conspicuous variation of the FWHM of the giant dipole resonance of cold nuclei with mass number (shell effects) testifies to the fact that the FWHM is essentially determined by $\Gamma^{\downarrow}$. This quantity is controlled by the coupling of the resonance to the nuclear surface and reflects both the static (inhomogeneous damping) and dynamic (collisional damping) deformation of the nuclear shape. Closed shell nuclei are spherical in average and less plastic than open shell nuclei. Heating up the system blurs, to a large extent this difference, as testified by the large values of $\langle \beta^2 \rangle$ observed also in coincidence of closed shells. This is because the nucleus behaves, under these circumstances, as a liquid drop, displaying large amplitude shape fluctuations which are thermally activated. Furthermore, the rapid rotation to which the compound nucleus is subject can induce conspicuous deformations in the system. As it will be shown in Chapters 9 and 10, large-amplitude thermal fluctuations and angular momentum splitting, that is, inhomogeneous damping (cf. Chs. 3 and 9), seem to account for essentially most of the changes observed in the FWHM of the giant dipole resonance (cf. Fig. 5.3(c)) as a function of temperature and angular momentum. These effects also seem to explain the anisotropy of the photons observed in the γ-decay of compound nuclei in the energy region of the giant dipole resonance, (cf. Sect. 6.1.2).

5.2 Experimental Techniques

In this section we shall discuss the experimental conditions to be met in order to be able to detect high-energy γ-rays emitted in fusion reactions induced by heavy-ion collisions.

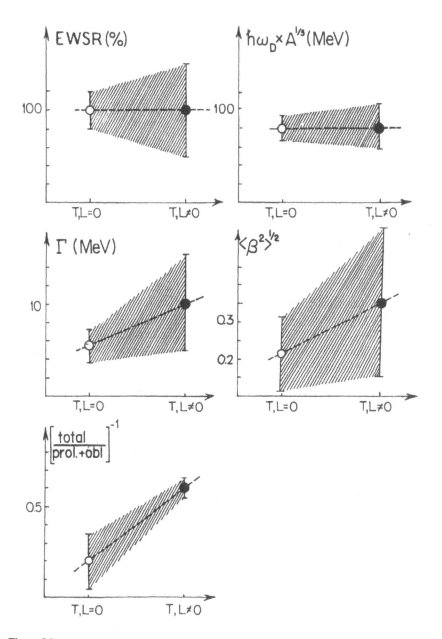

Figure 5.3. Average properties extracted from Figs. 2.5 and 5.2, of the giant dipole resonance based on the ground state (T,L = 0, open circles) and as observed in the γ-decay of intrinsically excited nuclei (T,L $\neq$ 0, black dots), averaged over mass, excitation and angular momentum. In (a) the strength of the GDR is given in unit of the EWSR. In (b) the centroid energy is displayed. In (c) the FWHM is shown, while in (d) the second moment of the quadrupole deformation parameter β is given. In (e) is displayed the normalized ratio of the cases where displays a prolate or an oblate configuration.

Because of the high density of states displayed by excited nuclei, and of the small probability with which the giant dipole resonance is present in thermal equilibrium with other degrees of freedom in the compound nucleus, efficiency, rather than energy resolution, has been the main concern in the design and construction of the experimental setups. Consequently, high-energy resolution γ-detectors, like Ge-crystals are seldom used as, in most cases, due to their small size, they display small efficiencies. On the other hand, if one could use very efficient γ-detector arrays with high-energy resolution (like e.g. EUROBALL, GAMMASPHERE, etc.) new possibilities could be opened in the study of giant resonances in hot nuclei. For example, one could study in detail the line shape of the giant dipole vibration based on superdeformed configurations and ask whether the low-energy component of the resonance displays features which go beyond those expected from the complexity of the nuclear spectrum at high energy (e.g. look for remnants of "doorway coupling," cf. Sect. 4.2.1). More in general, study the low-energy tail of the resonance, a quantity which plays a central role in determining the path followed by the γ-decay of hot nuclei. One could also ask whether the energy centroid of the resonance changes at all with temperature (cf. Sect. 10.5), and to which extent, such an eventual change affects the energy weighted sum rule limits, etc.

To date, mainly scintillator detectors, that is, detectors made of one of the inorganic materials, NaI, BaF_2 and BGO, have been used in the study of the giant dipole vibration in hot nuclei. The technique of converting the scintillation light produced by the interaction of gamma-rays with these materials in electric pulses is one of the oldest in the field of radiation measurements. However, it is only of recent date that one is able to produce scintillator crystals of very large volumes and also photomultiplier tubes (for converting the light pulse into an electron pulse) with large photocathode areas. Consequently, the availability of large and efficient detector arrays specifically constructed to measure high-energy gamma-rays is also of recent date. Examples of these arrays are presented in the next section.

The two most important requirements that were kept in mind in the design of these detector arrays were, the good gain stability and the good γ-neutron discrimination. This discrimination is usually made through time of flight techniques. In this connection, it has been shown particularly advantageous to employ detectors made out of BaF_2 crystals. This is because they produce light characterized by a fast scintillation component in the ultraviolet region, giving rise to very good time resolution, of the order of 0.8 ns, for large volume crystals. As a time reference, use is made of either a signal from the beam accelerator, or a signal from another reaction product of the fusion event.

The constancy in the proportionality between the electron pulse height and the energy deposited in the crystal by a photon, observed as a function of this energy, is measured by the so called "gain stability." It is unfortunate, within the present context, that photomultipliers coupled to scintillators do not have a gain stability that is adequate to the rapid variations of count rates typical of in-beam measurements, in which the γ-emission from the target is detected in the presence of an accelerated beam hitting the target. Ways to overcome this limitation are discussed below.

In the measurements of discrete γ-transitions, electronic drifts can be observed by the shift of the peak energies. On the other hand, in the detection of γ-transitions with continuous energy, the electronic drifts which can blur or even wash out spectral line shapes are difficult to see, since the spectrum is featureless. This problem is particularly serious for spectra with exponential shapes as the ones associated with the γ-decay of the giant dipole resonance from compound nuclei. In principle, this problem could be solved by making reference to the discrete part of γ-spectra of the compound nucleus decay. However, this complex spectrum is usually not recorded by the detector system employed to measure the giant dipole resonance γ-decay. This is because, as a rule, the energy resolution of the corresponding arrays is not adequate to recognize particular gamma-transitions. Furthermore, because the multiplicity of low-energy gamma-rays is 10^3 times larger than that of the high-energy gamma-rays of interest, it is advantageous, in the study of the GDR, to reject the first type of photons. This is achieved by setting an electronic threshold at around $E_\gamma = 5$ MeV. In addition, to reduce the pile-up of the high rate signals produced by low-energy γ-rays, together with those produced by high-energy γ-rays, a lead attenuator is placed on the front face of the detector.

Under such condition, the γ-spectrum associated with fusion reactions induced in heavy ion processes displays no narrow peak in the spectrum which can be used to monitor the gain stability. One way of solving this deficiency consists in introducing an artificial peak by direct light shining in the detector. This can be accomplished with the help of a light-emitting diode, coupled to appropriate electronics to tag these particular events. In the case of the BGO crystals, internal radioactivity leads to peaks with energy that is usually above the value of the electronic threshold chosen in the experiments, and which are used to monitor gain fluctuations.

The level of accuracy needed for the gain stabilization is of the order of 0.1–0.2%. For example, taking into account that the number of emitted photons decreases by an order of magnitude when the transition energy is increased by about of 5 MeV, a gain shift of 1% can change the amplitude of the angular distribution of the photons in the GDR region by about 10%.

Another technical requirement to be fulfilled in order to be able to compare the data with model calculations, is that the detector lineshape and efficiency (response function) must be known over the energy range of interest (5–30 MeV). The response function of the detector to the high-energy photons can be measured using photon beams. It can also be calculated simulating the propagation of the electromagnetic shower produced by the interaction of the high-energy γ-rays in the scintillator. Such calculations, based on Monte Carlo simulations, are usually carried out making use of the computer codes EGS and GEANT (Brun et al. (1986)). Results of typical calculations of the response function to monoenergetic γ-rays of detectors of the type used in the array HECTOR (cf. Sect. 5.3), are shown in Fig. 5.4. The calculations were

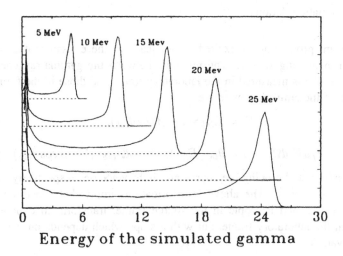

Energy of the simulated gamma

Figure 5.4. Calculated energy spectra produced by monoenergetic γ-rays interacting in a large volume (14.5 cm in diameter and 17 cm long) BaF_2 scintillator detector. The simulation of the electromagnetic shower produced by the interacting γ-rays in the detector was made using the code GEANT. (Adapted from Camera (1992).)

carried out making use of the program GEANT. A general feature of these calculations is that, as the energy of the γ-rays entering the detectors increases, the spectral distribution becomes broader, although the full energy peak is well defined for the region of interest. The calculations of the response functions are usually tested using measurements of high-energy monochromatic photon beams. The energy calibration of these detectors presents particular problems, since no simple radioactive source producing gamma-rays with energy similar to that of the GDR exists. Consequently, particular α-radioactive sources and nuclear reactions are employed for calibration purposes. For example, γ-rays with an energy of 6.13 MeV can be obtained with a $Pu^{13}C$ source in which the α particles from the spontaneous decay of Pu induce the reaction $\alpha + {}^{13}C \rightarrow {}^{16}O + n + \gamma$ (6.13 MeV). Because of neutron production, such sources have to be handled with particular care. Among the nuclear reactions used for calibration purposes, one can mention the compound nucleus reaction $p + {}^{11}B$ which populates the GDR in ${}^{12}C$ ($E_{GDR} = 22.5$ MeV), and which subsequently can decay to the ground state by emission of a single γ-ray with energy $E_\gamma = E_{CM} + Q$. The quantity E_{CM} is the energy in the center of mass of the $p + {}^{11}B$ system while Q denotes the Q-value for that reaction. Furthermore, a 15.1 MeV state in ${}^{12}C$ which decays to the ground state by a strong M1-transition can be populated by several different reactions. Often employed are also the reaction ${}^{2}H({}^{11}B, n\gamma){}^{12}C$ at a bombarding energy of 19 MeV and the ${}^{12}C(p, p'){}^{12}C$ reaction at a bombarding energy of 27 MeV. At higher energies, cosmic rays (relativistic muons), deposit energy in the crystal proportional to the traversed thickness and gives rise to a wide peak.

5.2.1 Kinematic relations

The in-beam production of excited nuclei leads to the emission of gamma-rays from a moving source. The relations between the gamma-ray energies and cross sections measured in the laboratory frame and those in the frame of reference of the emitting source are

$$E_s = E_{lab}\gamma(1 - \beta\cos\theta_{lab}), \qquad (5.4)$$

$$d^3\sigma(\theta_s, E_s)/dE_s d\Omega_s = d^3\sigma(\theta_{lab}, E_{lab})/dE_{lab} d\Omega_{lab}\gamma(1 - \beta\cos\theta_{lab}). \quad (5.5)$$

Here $\beta = v/c$, v and c being the source and light velocities respectively, while $\gamma = 1/(1 - \beta^2)^{1/2}$. The above transformations imply that spectra, which are exponential and isotropic in the source frame, transform to exponential spectra in the laboratory frame, but with a slope which depends on the angle of observation.

5.3 High-Efficiency Multidetector Arrays

In the present section we focus the discussion on the description of the multidetector systems constructed at the end of the 1980s and beginning of the 1990s and used in the study of giant dipole vibrations in hot nuclei. In particular we use as example the detector array HECTOR (Maj et al. (1994)) that was designed for this particular purpose. The much larger detection systems such as TAPS (Novotny et al. (1991)) and MEDEA (Migneco et al. (1992)) built to study meson production, bremsstrahlung radiation and nuclear fragmentation phenomena were eventually also employed for studies of giant dipole vibrations in nuclei at very high excitation energy.

The essential part of the HECTOR array[2] consists of 8 large volume scintillators of BaF_2 for the detection of high-energy photons and of 38 crystals of the same material and of smaller dimensions and with hexagonal shapes, which are used as a multiplicity filter for the low-energy γ-ray transitions. The geometry of such arrangement is schematically illustrated in Fig. 5.5. The large volume BaF_2 are positioned at different angles with respect to the beam direction and with the front face at 30 cm from the target center. The detectors of the multiplicity filter form two honey-comb structures, one above and one below the reaction plane.

The material BaF_2 was chosen because the scintillation light has a fast component leading to a good time spectra (600 ps in the present case). This

---

[2] The HECTOR array was built as a Danish-Italian collaboration and was completed in 1992. It was located, from the beginning of the construction which took place in 1989 at the Tandem Laboratory of the Niels Bohr Institute, at Risø, Denmark. It has continuously operated at this laboratory until 1997 when it will be moved to the Laboratory of Legnaro, a national Italian nuclear physics facility operated by the Istituto Nazionale di Fisica Nucleare (INFN).

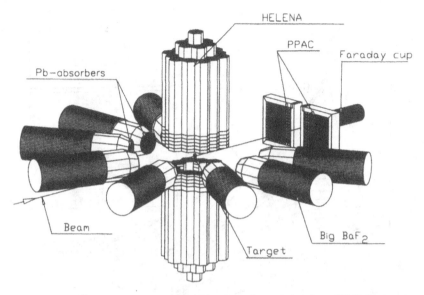

Figure 5.5. Schematic illustration of the high-energy photon spectrometer HECTOR. The heart of this detector array consists of 8 large scintillators of BaF_2 for the detection of photons produced in nuclear collisions with energy in the range 5–100 MeV. The setup also includes a multiplicity filter consisting of 38 smaller BaF_2 scintillators used to measure the angular momentum of the rotating nuclei produced in the nuclear reactions and a gas counter system (position sensitive PPAC) for the measurement of heavy charged fragments from such reactions.

property is useful in separating γ-rays from neutrons by time of flight, using rather short flight paths to which are associated large solid angles subtended by the detectors. To transmit and convert into an electric pulse the fast component of the detector scintillator light which lies in the ultraviolet (with wavelength of 220 nm), it is necessary to optically couple the crystals to photomultipliers with a quartz window on the photocathode and with very fast risetime in the anode signal. The high-energy detectors are equipped with a light emitting diode system which shines in the crystal, through an optic fiber, and which is thermally stabilized. The variation in the counting rate induces a gain drift that is measured by the centroid position of the peak produced by the direct light emitted by the diode. Such peak can be then isolated from the rest of the spectrum. With such a setup, it is possible to correct the gain changes down to the level of 0.1%.

Pile-up events, that is, events involving two γ-rays or a neutron and a γ-ray in which a γ-ray or the neutron interacts in the detector before the signal of the other γ-ray is completely processed, can also be identified. In fact, the pile-up of gamma rays produces electron pulses that are characterized by a ratio of the charge contained in the fast part of the pulse and in the entire pulse that is different from that associated with non-pileup events. The effect is illustrated in Fig. 5.6 where the charge contained in the first part of the signal, indicated by the gate named "fast," and that of the entire signal, named

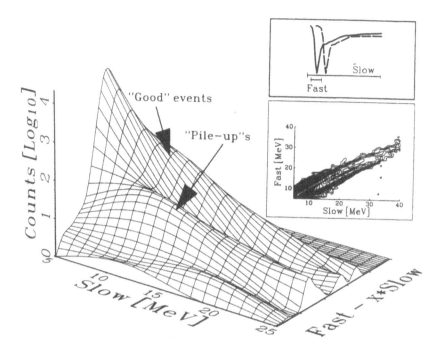

Figure 5.6. Illustration of pulse-shape discrimination obtained by applying two gates of different widths to the energy pulse of a BaF$_2$ detector and integrating the charge in each window (see top inset). When the detectors are operated in air, the contribution from charged particles is generally modest and the main purpose of the pulse-shape discrimination is to reduce pile-up. The lower inset shows a projection of the three-dimensional distribution. (Adapted from A. Maj et al. (1994).)

"slow" gate, are plotted together for both pile-up and normal events. Making use of a charge sensitive analog-to-digital converter (ADC), the fast component is measured integrating the current signal from the anode over a narrow time interval (30 ns) while the slow component is obtained integrating over a wider time interval (1 μs). An important feature of the analog to digital converters that were employed in the experiments, is their capability of correcting for any charge present between two consecutive events, so that the energy calibration curve has zero energy in channel zero. The technique described above for discriminating pile-up events from normal events, is the same as that used to identify the variety of charged particles emitted in experiments aimed at measuring particle spectra and the associated multiplicities in the study of heavy ion reaction mechanisms. At variance to the HECTOR setup, in the case of particle detection, the scintillators have to be located in a vacuum chamber.

Standard electronics was employed for the HECTOR detector system. It was chosen so it could provide, as basic trigger, a coincidence between one of the large-volume BaF$_2$ detectors with one of the small-volume BaF$_2$ detectors of the multiplicity filter. To measure the time of flight from the target, for the events producing electric signals above the electronic detection threshold, all the anode signals were sent as inputs to a unit that serves as a constant fraction discriminator (CFD) modulus. In order to measure the coincidence fold and the sum energy in the small-volume BaF_2 detectors, the associated logic and analogic signals were summed and then converted with two ADC's.

The efficiency, the multiple hit probability, the cross talk and the response function providing the conversion of each coincidence fold with the distribution of gamma-rays emitted by compound nuclei, was determined with a method based on coincidence data of radioactive sources (Maj et al. (1994)). The conversion from γ-ray number (multiplicity) M_γ, into spin value J, is based on statistical-model arguments (see the next section) and is given by the relation,

$$J = 2 \times M_\gamma + C. \tag{5.6}$$

Here M_γ is the multiplicity of rotational-decay γ-ray transitions, assuming each γ-ray to carry two units of angular momentum. The quantity C is a correction factor taking into account the angular momentum removed by particle evaporation and by statistical γ-ray transitions. It can be calculated using the statistical model (Hauser and Feshbach (1952)), and the resulting value depends on the compound nucleus reaction considered.

5.4 Statistical Model of γ-Decay

At variance with γ-spectroscopy of discrete levels, where the measured cross sections and lifetimes are the end results of the measurement, in that they can be directly compared with the predictions of nuclear structure models, continuous γ-spectra emitted from hot nuclei as shown in Fig. 5.1 can yield equivalent information, only after a statistical model analysis of the experimental data has been carried out.

The basic assumption made in such analysis, is that all nuclear degrees of freedom have reached statistical equilibrium before the cooling process starts (compound nucleus). The assumption is also made of the validity of detailed balance (Hauser and Feshbach (1952)). That is, that the transition matrix elements associated with the formation of the compound nucleus $a + A \rightarrow C$, are the same than those associated with the decay process $C \rightarrow a + A$. Consequently, the decay rates may be determined from the cross section of the inverse excitation process[3].

The width $d\Gamma_x(e_x)/de_x$ per unit energy for the decay of nucleus 1 at excitation energy E_1, spin J_1 and parity π_1 to nucleus 2 at excitation energy E_2,

---

[3] The notation used here is the same as that used in Pühlhofer (1977), in connection with the description of the computer code CASCADE

spin J_2 and parity π_2 by emission of particle x is given by

$$d\Gamma(e_x)/de_x = \frac{\rho_2(E_2, J_2, \pi_2)}{2\pi\rho_1(E_1, J_1, \pi_1)} \Sigma^{J_2+s_x}_{S=|J_2-s_x|} \Sigma^{J_1+S}_{L=|J_1-S|} T^x_L(e_x). \qquad (5.7)$$

Here $e_x = E_1 - E_2 - B_x$ is the kinetic energy of particle x, B_x and s_x its separation energy and spin, respectively, and L its orbital angular momentum. The particle transmission coefficients $T^x_L(e_x)$ for the scattering of x by the nucleus 2 are determined from averaged optical model parameters (Rapaport at al. (1979)).

To write the expression of the decay width for statistical E1 γ-decay, one first assumes that (Axel–Brink hypothesis, cf. Brink (1955)) every nuclear level has a giant dipole resonance built on it (cf. Fig. 1.2). Using the reciprocity theorem (detailed balance), one obtains for the γ-decay width

$$d\Gamma^\gamma(E_\gamma)/dE_\gamma = \frac{\rho_1(E_2, J_2, \pi_2)}{\rho_1(E_1, J_1, \pi_1)} \frac{\sigma_{abs}(E_\gamma)}{3} \frac{E_\gamma^2}{(\pi\hbar c)^2}, \qquad (5.8)$$

where $E_\gamma = E_1 - E_2$, $J_2 = J_1$ or $J_1 \pm 1$. In the last two equations, the usual parity restriction, that is $\pi_1 \cdot \pi_2 = -1$ for electric dipole transitions is applied. The quantity $\sigma_{abs}(E_\gamma)$ represents the averaged absorption cross section assumed to have at finite temperature T, the same form as the one measured at T = 0 MeV, namely a single or double-Lorentzian shape,

$$\sigma_{abs}(E_\gamma) = \frac{4\pi e^2 \hbar}{Mc} \frac{NZ}{A} \Sigma^2_{j=1} \frac{S_j \Gamma_j E_\gamma^2}{(E_\gamma^2 - E_{Dj}^2)^2 + E_\gamma^2 \Gamma_j^2}. \qquad (5.9)$$

Here S_j, E_D and Γ_j are the oscillator strength, centroid and the width of the different components of the giant dipole resonance line shape. The fusion cross section leading to compound nucleus formation σ_c, is commonly calculated by approximating the entrance channel transmission coefficients by a Fermi function with a smooth cutoff, that is

$$\sigma_c = \Sigma_L \sigma(L) = \pi\lambda/2\pi\Sigma_L(2L+1)T_L(E), \qquad (5.10)$$

where $T_L = (1 + \exp((L - L_0)/d))^{-1}$, with L_0 chosen to match measured fusion cross sections. That is, and assuming a triangular distribution for $\sigma(L)$, L_0 is the maximum angular momentum at which the compound nucleus has been formed and $L_{mean} \approx 2L_0/3$ is the mean L-value of the compound system. The expression used for the level density is

$$\rho(U, J) = \frac{a^{1/2}(2J+1)}{12\theta^{3/2}(U+T)^2} \exp(2\sqrt{aU}), \qquad (5.11)$$

where the intrinsic excitation energy is defined as

$$U = E^* - \Delta(T) - E_{rot}, \qquad (5.12)$$

and where

$$E_{rot} = \frac{\hbar^2}{2\Im_{rig}} J(J+1). \qquad (5.13)$$

In the previous expressions, E^* denotes the excitation energy, a the density of levels parameter, $\Delta(T)$ the temperature dependent pairing energy, J the total angular momentum of the system and T the nuclear temperature. The quantity $\Im_{rig}$ is the rigid moment of inertia. The above expression of the level density takes into account the fact that the energy which is tied up in a single degree of freedom, like e.g. rotational motion, cannot be used in the thermalization process. This is the reason why U is known as the heat or intrinsic excitation energy (cf. also Sect. 11.5.1). On the other hand, the members of the rotational bands contribute to the total number of levels, as testified by the pre-factor in Eq. (5.11).

Since the parameters S_j, E_D and Γ_j characterizing the giant dipole resonance are extracted from a statistical model analysis of the measured γ-spectra, analysis which depends sensitively on $\rho(E, J)$, an accurate knowledge of the level density parameter a, at different values of T, and eventually J, is needed. This requirement is in general difficult to fulfill, and constitutes a serious source of uncertainty in the extraction of nuclear structure information from compound nucleus decay, not only in the present case, but also in the case of the study of rotational motion in warm nuclei (Ch. 11).

In the case where the statistical analysis is carried out making use of the code CASCADE, the value of the level density parameter is dealt with in the following way. For $E^* \leq$ 8–10 MeV the value of a is taken from the work of Dilg et al. (1973), based on experimental spectra at very low energy and (n, γ) data near the neutron binding energy. For $E^* \geq$ 15–20 MeV, the validity of the liquid drop regime where neighbouring nuclei are expected to behave similarly, is assumed. In this case a is chosen equal to $A/8$ MeV^{-1}, A being the compound nucleus mass number. In addition, in the case of the analysis of a spectrum associated with the decay of a highly excited ($E^* \geq$ 150 MeV) nucleus, the a parameter is varied at each step of the deexcitation cascade. In fact, studies of energy and multiplicity of charged particles emitted following compound nucleus reactions seem to indicate a value of a = A/8 MeV^{-1} for excitation energies lower or equal to 1 MeV per nucleon (Nebbia et al. (1986)). However, to fit the multiplicity of charge particles emitted by compound nuclei with larger excitation energies, a value of a= A/12 MeV^{-1} seems to be required.

The calculation of the total γ-spectra is made by summing all the contributions along the deexcitation chain, starting from the original compound nucleus, and following the emission of the daughter nuclei populated by particle emission. Consequently, a high-energy γ-ray spectrum does not reflect neither a single temperature, nor a given mass number, but a weighted average over all the possible values of T and A.

A typical temperature decomposition of a total γ-spectrum emitted in the entire deexcitation process by a nucleus of mass A = 100 formed at an initial temperature T = 2 MeV, is shown in Fig. 5.7. From this figure one can see that the contributions to the total spectrum, from the different temperature regions vary with γ-ray energy. This variation has consequences concerning

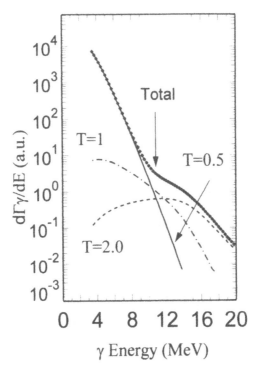

Figure 5.7. A calculated high-energy γ-ray spectrum as the best fit to experimental data for ^{110}Sn at $E^* = 92$ MeV. Some of the contributions to the total spectrum at different level of the decay chain are shown. The yield of γ-rays emitted in the first step of the decay is the one labeled with T = 2 (dashed line), while the gamma yield from the colder daughter nuclei at T = 1 and at T = 0.5 MeV are shown with dashed-dotted and full drawn lines, respectively. (Adapted from Camera (1995).)

the type of decays taking place at the different stages of the cooling process. In particular, γ-decay with $E_\gamma \geq B_n$, B_n being the neutron separation energy, competes directly with particle decay and is favored to occur early in the cooling process. For $E_\gamma \leq B_n$, the gamma-decay occurs primarily after the excited nucleus has cooled down to excitation energies lying to within a value of the order of one-neutron separation as measured from the yrast line.

Because most of the initial angular momentum is carried away by stretched E2-transitions, that is, transitions each carrying two units of angular momentum and occurring late in the cooling process, the γ-decay of the giant dipole resonance takes place from a nucleus displaying essentially the distribution of angular momenta, and therefore of temperature (cf. Eq. (5.12)), of the original compound nucleus. This effect is a source of uncertainty in extracting the parameters of the resonance from the statistical analysis of the experimental data.

5.4.1 Simple estimates

To gain physical insight into the decay process of high-energy gamma-rays, it is useful to derive analytical expressions for the particle- and γ-decay widths, making use of simplifying assumptions. It can be shown that, the summation over spin and orbital angular momentum together with the integral over kinetic energy entering in the calculation of the neutron-decay width, is essentially proportional to T^2. This result, combined with a first order expansion of the level density in T, leads to the following expression for the ratio of the γ-decay probability per unit energy to the total probability for neutron emission,

$$\frac{d\Gamma(E_\gamma)/dE_\gamma}{\Gamma_n} \sim \frac{E_\gamma^2}{T^2} \exp(\frac{B_n - E_\gamma}{T}) \frac{E_\gamma^2 \Gamma}{(E_\gamma^2 - E_D^2)^2 + E_\gamma^2 \Gamma^2}. \tag{5.14}$$

In deriving this expression, the E1-absorption cross section has been assumed to be described by a Lorentzian function of width Γ and centroid E_D. The quantity $\Gamma_n = \int d\Gamma_n(e_n)$ is the total neutron width. Many of the features of statistical gamma-decay can be deduced from the above expression.

For $E_\gamma \leq B_n + 2T$ (where $2T$ is approximately the kinetic energy carried away by neutrons, T being the temperature of the system), the γ-decay probability increases with decreasing T. Consequently, most of the associated gamma-rays originate near the end of the decay cascade. For $E_\gamma \geq B_n + 2T$ the opposite is true. In other words, energetic γ-decay occurs as a rule before particle emission, reflecting the hot part of the decay cascade. In discussing the competition between charged particles and γ-rays, the quantity B_n in the above inequality is substituted by the single-particle separation energy plus the Coulomb barrier energy.

The effective temperature associated with the state of spin I upon which the GDR is built (cf. Fig. 1.3), that is, the state populated by γ-decay is given by (cf. Eq. (7.18))

$$T = \left(\frac{E^* - E_D - E_{rot}}{a} \right)^{1/2}. \tag{5.15}$$

In the above expression the quantity E^* is the excitation energy of the nucleus, E_D the energy centroid of the giant dipole resonance, E_{rot} the energy of the yrast band at spin I (cf. Eq. (5.13)) and a the level density parameter. The rotational energy may be estimated in terms of the mean angular momentum $\langle J \rangle \approx L_{mean}$ involved in the gamma-decay region under study.

Another useful quantity in the study of the decay of the compound nucleus is the branching ratio for high-energy γ-decay. At zero temperature, the γ-width for decay of a state J_1 to a state J_2 summed over the magnetic states M_1 is given by $\Gamma_\gamma = 5.2E_\gamma$ (MeV^2) NZ/A (eV). Since the total width of the GDR based on the ground state displays values (cf. Fig. 2.5(c)) ranging from 4 MeV (spherical nuclei) to 8 MeV (deformed nuclei), the ground state γ-branching-ratios Γ_γ/Γ are in the range 10^{-3} to 10^{-2}. The γ-branching

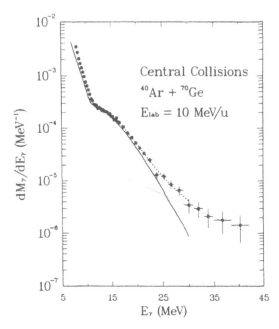

Figure 5.8. Measured γ-ray spectrum from the compound nucleus reaction ^{40}Ar + ^{70}Ge → ^{110}Sn at the bombarding energy 10 MeV/u. The dotted line shows the contribution from pre-equilibrium γ-rays due to the nucleon–nucleon collisions. Gamma-rays from the decay of the giant dipole resonance in the excited nucleus are concentrated in the range E_γ = 10–25 MeV. The full drawn line shows a spectrum calculated with the statistical model as described in the text. (Adapted from Bracco et al. (1989a).)

ratio for the decay to excited states (hot nuclei), can be deduced integrating the γ-decay strength for the first step of the γ-decay cascade and calculating the total width for the decay. These values are in general smaller (within an order of magnitude) than those corresponding to the T = 0 MeV decay.

An example of statistical model analysis of a measured high-energy γ-spectrum is given in Fig. 5.8. Central collisions in the reaction ^{40}Ar + ^{70}Ge at bombarding energy 10 MeV/u lead to fusion reactions and thus to a compound system, namely ^{110}Sn with excitation energy $E^* \approx 230$ MeV which, eventually, cools down by particle- and γ-emission. The corresponding spectrum has been corrected for Doppler shift assuming emission from a system moving with the center-of-mass velocity. It has also been corrected for the detector response function. Also shown in Fig. 5.8, is a statistical model calculation obtained making use of the code CASCADE. In this calculation the γ-ray emitted at all steps of the deexcitation chain of the (^{110}Sn) compound nucleus were computed. The dipole strength function was assumed to have a Lorentzian shape with a centroid $E_D = 16$ MeV, a width $\Gamma_D = 13$ MeV, and to exhaust 100% of the energy weighted sum rule. In this calculation, E_D and Γ_D were assumed to be constant along the deexcitation process. In addition,

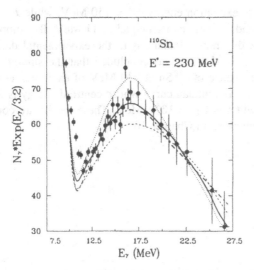

Figure 5.9. γ-ray energy spectrum of Fig. 5.8 from which the bremsstrahlung component has been subtracted in comparison with statistical model calculations. All spectra are multiplied by the function $\exp(E_\gamma/3.2)$. The statistical model calculations correspond to different values of the level density parameters and of the giant dipole resonance parameters as described in the text. (Adapted from Bracco et al. (1989a).)

a constant level-density parameter $a = A/8$ MeV^{-1} was used in the calculations. For this value of the level density parameter, the GDR spectra have been consistently described over a broad mass range at moderate excitation energies (up to ≈ 100 MeV).

To carry out a detailed comparison between the experimental findings and the statistical model calculations in the region of the giant dipole resonance, it is convenient to display the results in such a way that level density effects are basically eliminated. A way to achieve this is to multiply the spectra by an exponential function of the γ-ray energy E_γ. In the particular case of the data of Fig. 5.8 one has also to subtract the contribution due to the nucleon–nucleon bremsstrahlung radiation originating in the early stage of the collision, before the compound nucleus is formed. The bremsstrahlung component of the spectrum (shown with the dotted line in Fig. 5.8) is characterized by γ-energies greater than 25 MeV and starts to be important for bombarding energies of the order of 10 MeV per nucleon or larger. In Fig. 5.9 the spectrum obtained multiplying the data by the function $\exp(E_\gamma/3.2)$ is shown together with statistical model calculations multiplied by the same exponential function.

To learn about the sensitivity of the results on the parameters entering the model, the calculations were repeated for different values of these quantities. The full drawn curve was obtained using $a = A/8$ MeV^{-1}, $E_D = 16$ MeV and $\Gamma_D = 13$ MeV. The dashed-and-dotted curves correspond to $\Gamma_D = 15$ MeV and $\Gamma_D = 11$ MeV, respectively. The dot-dashed curve corresponds to a calculation in which the level density parameter was decreased to a value $a =$

A/12 MeV^{-1} above excitation energy $E^* = 150$ MeV, while E_D was chosen equal to 15 MeV and Γ_D was chosen equal to 11 MeV. By comparing, on an absolute scale, the different calculations to the experimental data within the range $11 \leq E_\gamma \leq 25$ MeV, it was concluded that the strength function of the giant dipole resonance of ^{110}Sn at 230 MeV of excitation energy, can be described by a single Lorentzian curve whose centroid energy is $E_D = 16\pm 1$ MeV and whose width is $\Gamma_D = 13^{+1}_{-2}$ MeV. These results are compared with the theoretical predictions in Chapter 10.

6

Dipole Oscillations: Experiment

In trying to review the main features extracted from the variety of experimental data which has been accumulated on the decay of the giant dipole resonance from excited nuclei, it is practical to divide them in two main parts: the first concerns the decay of compound nuclei with moderate excitation energies ($E^*/u < 1$ MeV (Sect. 6.1), while the second corresponds to the decay of highly excited nuclei (Sect. 6.2).

6.1 Moderate Excitation Energies

The detailed study of the γ-decay from compound nuclei with moderate excitation energies allows to follow the evolution of the nuclear structure from a region close to the yrast line, where the shell structure dominates, up to the region where nuclei can be described with the help of concepts of classical and statistical physics. In developing this subject, we shall concentrate our attention on the question of the role played by deformation and shape fluctuations, on the width of the giant dipole resonance. Examples from different regions of the mass table will be reviewed, namely light nuclei with A $\approx$ 40, medium mass nuclei with A $\approx$ 110 and medium heavy mass nuclei with A $\approx$ 170.

6.1.1 A $\approx$ 40

The nucleus ^{45}Sc was studied at excitation energies $E^* = 50$, 66.6, 76.7 and 88.9 MeV making use of the heavy ion fusion reaction ^{18}O + ^{27}Al at varying bombarding energies. The average angular momenta of the associated compound nuclei are $\approx$ 13, 18.5, 21.4 and 23.5 $\hbar$, respectively. The measured spectra, shown in the top row of Fig. 6.1 were analyzed making use of statistical model calculations in which the GDR strength function was parametrized in terms of two Lorentzian curves. The main results obtained from such an analysis were: a) the summed value of the widths of the two

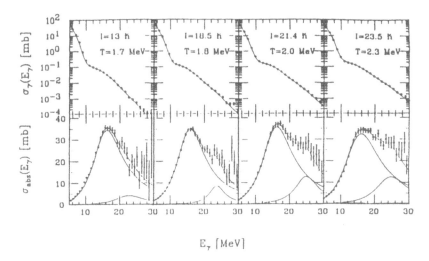

Figure 6.1. Top row: high-energy γ-spectra associated with the decay of the compound nucleus ^{45}Sc, at different temperatures and angular momenta. The full drawn lines are the results of statistical model calculations. Bottom row: data and calculations corresponding to the top row in linear scale as obtained with the prescription described in the text. (Adapted from Kicińska-Habior et al. (1993)).

Lorentzian components is larger than the FWHM of the resonance observed in the photoabsorption reaction, b) the energy splitting of the centroid of the two peaks implies sizable nuclear deformations. In connection with point b), the values of the quadrupole deformation parameter β range from a value of 0.35 at an excitation energy $E^* = 50$ MeV and average angular momentum 13 $\hbar$, to $\beta \approx 0.46$ at $E^* = 88.9$ MeV and average angular momentum 23.5 $\hbar$. This result is better seen displaying the spectra in a linear scale so that the region of the giant dipole resonance is emphasized. For this purpose, the quantity $F(E_\gamma) * Y_\gamma^{exp}(E_\gamma) / Y_\gamma^{cal}(E_\gamma)$ was extracted, and plotted in the bottom row of Fig. 6.1. In the above expression, the experimental spectrum is indicated by $Y_\gamma^{exp}(E_\gamma)$, while $Y_\gamma^{cal}(E_\gamma)$ is the best fit calculated spectrum, corresponding to the Lorentzian functions $F(E_\gamma)$.

The theoretical analysis of the data (cf. Alhassid (1994), Snover (1993)) seems to indicate that the system, not only becomes progressively more deformed as a function of angular momentum, but that it undergoes a transition from oblate to triaxial (approximately prolate) shapes. This shape phase transition of nuclei is similar to that taking place in rotating stars and described in terms of Jacobi and McLaurin's ellipsoids (Chandrasekhar (1969)).

6.1.2 A ≈ 110

Systematic information on the giant dipole resonance strength function is available for ^{108,112}Sn, over a broad range of excitation energies and angular mo-

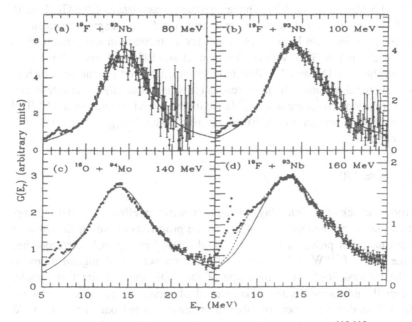

Figure 6.2. γ-ray spectra from the decay of the giant dipole resonance in ^{110,112}Sn isotopes at different excitation energies displayed in a linear scale (see text). The continuous curves show the results of best fits of statistical model calculations. (Adapted from Chakrabarty et al. (1987).)

menta. At variance with the previous case, in all experiments contributing to the Sn-systematics, the high-energy γ-rays were assigned to compound decays with the help of more exclusive measurements. In fact, the compound nucleus formation was identified by measuring, in coincidence, high-energy γ-rays with either the multiplicity of low-energy γ-rays of the evaporation residues, or by detecting the recoiling residues.

Results of measurements made employing the heavy ion reaction ^{19}F$+^{93}$Nb at bombarding energies 80, 100 and 160 MeV and ^{16}O $+$ ^{94}Mb leading to Sn-compound nuclei are shown in Fig. 6.2. A single Lorentzian-component-strength-function was used to reproduce the Sn-data. This implies that, either the nucleus has a spherical shape, or that thermal shape fluctuations are averaging out possible static deformations (cf. below as well as Ch. 10). An increase of the damping width of the giant dipole resonance from 5.8 MeV to 10.8 MeV is observed, for changes in the excitation energy from 50 MeV to 130 MeV and of the average angular momentum from ≈ 15 $\hbar$ to ≈ 40 $\hbar$.

In trying to disentangle the temperature and angular momentum effects on the FWHM of the GDR, measurements were carried out at fixed excitation energy for different values of the angular momentum, and for different temperatures at essentially fixed (and low) angular momentum. The FWHM of the GDR was found (Bracco et al. (1995)) to increase by ≈ 20% when the nuclear angular momentum increased from 40 $\hbar$ to 54 $\hbar$ at a fixed temperature

of T = 1.8 MeV (cf. Fig. 6.7). The temperature dependence of the GDR width
of ^{120}Sn (as well as that of ^{208}Pb) at approximately zero angular momentum,
was found to increase by a factor of ≈ 2 for an increase in temperature from
T$\approx$ 1 MeV to T $\approx$ 3 MeV (Ramakrishnan et al. (1996); cf. Sect. 10.6.1).

Both the A = 110 and 120 data have provided a solid ground on which to
test theoretical models. The main results (cf. Chs. 9 and 10) of these tests are:
a) deformation effects are responsible for the measured increase of the FWHM
of the GDR in hot Sn nuclei, b) the intrinsic width $\Gamma^{\downarrow}_{GDR}$ is independent of
temperature.

6.1.3 A $\approx$ 170

A further check of the interplay of angular momentum effects, liquid drop properties, and thermal shape fluctuations on the properties of the dipole vibration
in hot nuclei is provided by the results displayed in Fig. 6.3, corresponding
to the GDR of ^{176}W at $E^* = 70$ MeV and for a number of angular momenta
in the interval 36–55 $\hbar$. The shape of the GDR strength function remains
essentially unchanged within this interval of rotational frequencies (0.4–0.6
MeV). The difference between the observations carried out in Sn and in W
is well accounted for by the theory (cf. also Ch. 10), and is associated with
the fact that in the case of the Sn-experiments, the corresponding rotational
frequencies are much larger (≈ 1 MeV) than in the case of ^{176}W.

A compound nucleus at high angular momentum displays strong spatial
alignment, its angular momentum being essentially perpendicular to the reaction plane thus displaying zero projection on the beam axis. Under such
circumstances, it is possible to measure the angular distribution of the emitted
gamma-rays. The corresponding expression can be written in a compact way
choosing as the axis of quantization the direction of the total angular momentum of the nucleus, and making the assumption that $I \gg 1$ (semiclassical
approximation).

The giant dipole oscillation can be decomposed into vibrations parallel and
perpendicular to the nuclear angular momentum axis. The transition associated with the first type of vibration is of non-stretched character, that is, a
transition whose angular momentum is perpendicular to the nuclear angular
momentum I. The transition associated with the second type of vibration is of
stretched type. To further clarify these points, let us consider an axially symmetric nucleus with a prolate deformation rotating collectively. In this case
the angular momentum I of the system is perpendicular to the symmetry axis.
The dipole vibration along the symmetry axis corresponds to the low-energy
component of the GDR and has an angular momentum which is parallel to
that of the rotating nucleus. In keeping with the definitions given above, the
associated γ-transition is stretched. The two degenerate "high-energy" components are associated with vibrations along the short axes, and are parallel and
perpendicular to the nuclear angular momentum. Therefore they give rise to a

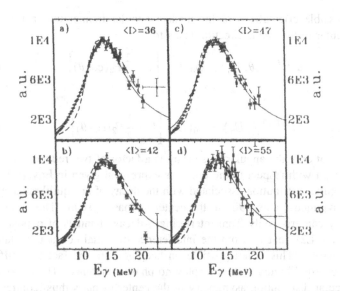

Figure 6.3. Gamma-ray spectra from the decay of the giant dipole resonance of ^{176}W at fixed excitation energy and for different values of the nuclear angular momentum. The data and the best fitting statistical model calculations are shown in a linear scale (see text). The obtained values of the giant dipole resonance width are approximately the same for the different angular momenta. The dashed lines are the results of calculations made with the model of thermal fluctuations in the adiabatic limit (cf. Sect. 10.8). (Adapted from Mattiuzzi et al. (1995).)

stretched and unstretched transitions, respectively. The angular distribution of γ-rays with respect to the direction of the angular momentum of the compound nucleus is, in the case of a stretched transition, given by the expression,

$$\omega^{|\Delta I|=1}(\theta_s, E_\gamma) = \omega_0(E_\gamma) \left[1 + \frac{1}{2}P_2(\cos\theta_s)\right], \qquad (6.1)$$

which in the case of a non-stretched transition becomes

$$\omega^{\Delta I=0}(\theta_s, E_\gamma) = \omega_0(E_\gamma) \left[1 - P_2(\cos\theta_s)\right]. \qquad (6.2)$$

In the two previous expressions, $\omega_0(E_\gamma)$ is the angular independent part of the cross section, $P_2(\cos\theta_s)$ is a Legendre polynomial which depends on the angle θ_s between the emitted GDR photon and the direction of the angular momentum of the emitting nucleus. Typical experiments in which the direction of the nuclear angular momentum can be determined, are those in which the compound nucleus formed in a fusion reaction, fissions. By measuring the position of the fission products, the direction of the nuclear angular momentum can be obtained, being perpendicular to the plane defined by the momenta of the fission fragments (semiclassical approximation).

When the angular distribution is measured with respect to the beam direction, one has to remember that the angular momentum of the nucleus can have any direction within the plane perpendicular to the beam. Averaging over all the possible orientations of the nuclear angular momentum in the plane perpendicular to the beam direction leads to

$$\omega^{|\Delta I|=1}(\theta, E_\gamma) = \omega_0(E_\gamma)\left[1 - \frac{1}{4}P_2(\cos\theta)\right], \tag{6.3}$$

and

$$\omega^{\Delta I=0}(\theta, E_\gamma) = \omega_0(E_\gamma)\left[1 + \frac{1}{2}P_2(\cos\theta)\right]. \tag{6.4}$$

Note that not only the angular anisotropy is a factor of two reduced but it has opposite sign with respect to the previous expression given in Eqs. (6.1–6.2).

The angular distribution associated with the decay of an equilibrated system must be symmetric about 90° in the center of mass system. This is because asymmetry can arise only from interfering γ-decay channels of opposite parity (e.g., E1–E2), that is, from the interference of initial compound states of different parity. This is not possible, in keeping with the fact that different states of an equilibrated system display no phase coherence. The observation of an angular distribution asymmetry in the center of mass thus requires both the presence of interfering radiation of opposite parity plus a pre-equilibrium reaction mechanism with phase coherence between different reaction amplitudes.

The anisotropies of the angular distribution depend on the deformation of the nucleus and on the orientation the system has with respect to the axis of rotation. The example of axially symmetric shapes is illustrated in Fig. 6.4. In this figure, the angular distribution coefficient $a_2(E_\gamma)$ relative to the beam direction is shown. This coefficient is calculated making use of the GDR absorption cross section distributions corresponding to stretched transitions σ_{-1} and σ_{+1} and that associated with the unstretched transition σ_0 according to

$$a_2(E_\gamma) = \frac{\sigma_0 - 0.5(\sigma_{-1} + \sigma_{+1})}{4(\sigma_0 + \sigma_{+1} + \sigma_{-1})}. \tag{6.5}$$

The experimental values of $a_2(E_\gamma)$ are obtained by fitting the data with the function $W(\theta, E_\gamma) = W_0[1 + a_2(E_\gamma)P_2(\cos\theta)]$.

In Fig. 6.4 the convention used to measure the quadrupole shape parameters β and γ is: (a) $\gamma = 0$ (120°) for prolate shapes rotating collectively (non-collectively), namely with the nuclear angular momentum perpendicular (parallel) to the nuclear symmetry axis, (b) $\gamma = +60°$ ($-60°$) for oblate shapes rotating collectively (non-collectively). It is then apparent that the strength functions depend only on the type of deformation, namely whether the system is prolate or oblate while the angular distribution coefficient $a_2(E_\gamma)$ depends also on the nuclear orientation of the system (collective-or non-collective type of rotation). This is because the overlap of the stretched and unstretched components of the GDR depends on the nuclear orientation. The picture discussed above is only correct in the semiclassical limit. In fact, when the rotational

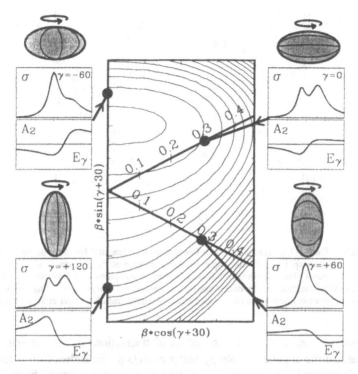

Figure 6.4. Schematic illustration of the giant dipole resonance strength function and angular distribution as a function of γ-ray energy for nuclei of a fixed deformation ($\beta = 0.3$) but of shapes of different types. (Adapted from A. Maj et al. (1994).)

frequency of the nucleus is not very high ($\hbar\omega_{rot} \leq 0.5$ MeV), orientation fluctuations become important. The associated loss of alignment changes the overlap of the various vibrational components, and the angular anisotropy is attenuated.

All measurements of the angular distribution in the GDR energy region made to date, are consistent with nuclear prolate shapes rotating collectively and with oblate shapes rotating non-collectively. Typical $a_2(E_\gamma)$ coefficients are shown in Fig. 6.5. They were measured in the γ-decay of the compound nucleus ^{176}W at a temperature of T = 1.4 MeV and for different values of the nuclear angular momentum. It is seen that the angular anisotropy around E_γ = 10–12 MeV increases by more than a factor of two when the angular momentum increases by 50%. This behaviour is in contrast with the observed near constancy of the GDR width, a constancy which can be explained in terms of the thermal fluctuation of the nuclear orientation relative to the nuclear angular momentum. The wobbling of the nucleus around the nuclear angular momentum axis becomes conspicuous at low rotational frequencies ($\hbar\omega_{rot} \approx 0.2$–0.3 MeV). In fact, under these circumstances, a deformed nucleus may appear to be almost spherical, that is, display very small values of $a_2(E_\gamma)$. When the

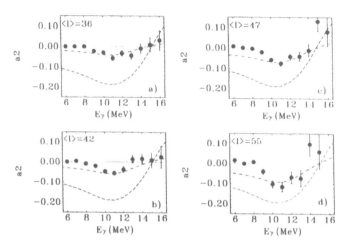

Figure 6.5. The $a_2(E_\gamma)$ coefficients as a function of E_γ measured for the γ-decay of ^{176}W. The four plots correspond to different spin windows centered at the average value (in $\hbar$ units) shown in each panel. The experimental points (full dots) are compared with the results of adiabatic calculations which include only thermal shape fluctuations (dashed line) and including both thermal shape and orientation fluctuations (dot-dashed line). (Adapted from Mattiuzzi et al. (1995).)

rotational frequency increases, the effect of fluctuations in the orientation is reduced and the angular anisotropy better reflects its dependence on deformation. Theoretical calculations, which take into account thermal fluctuations in both the shape and the orientation degrees of freedom, provide an overall account of the experimental findings. This can be seen from Fig. 6.5, where the theoretical results (dashed-dotted curve) are compared with the data. From the same figure one can also see that calculations which only include shape fluctuations fail to reproduce the data (cf. also Ch. 10).

Measurements of angular distributions and of strength functions were also made in the case of the nuclei ^{109,110}Sn at T = 1.8 MeV, and for different values of the angular momentum. The combined analysis of the observed $a_2(E_\gamma)$ values and of the strength function proved to be instrumental in assessing the role played by the different mechanisms contributing to the damping of the GDR.

Before discussing the data, let us illustrate with a simple example the relevance of studying simultaneously the GDR line shape and the $a_2(E_\gamma)$-coefficient (cf. Fig. 6.6). In this example we have considered an increase of the FWHM of the resonance of 1 MeV. If this increase is due to an increase of the collisional damping width (dotted-dashed curve in the top row of Fig. 6.6) the associated $a_2(E_\gamma)$ is almost unchanged. If, on the contrary, the FWHM

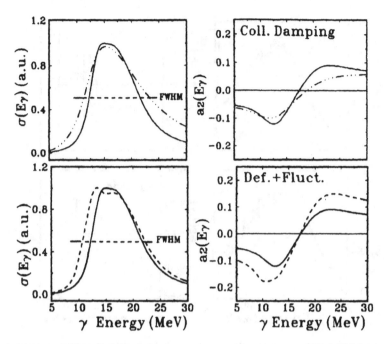

Figure 6.6. Strength function (left) and angular distribution coefficient $a_2(E_\gamma)$ (right) calculated with the adiabatic thermal fluctuation model at T = 2 MeV. The collisional damping width of the resonance was parametrized according to $\Gamma_D^\downarrow = \Gamma_0^\downarrow (E/E_D)^{1.9}$ (cf. Carlos et al. (1974)). In the calculations displayed at the top of the figure, the so-called intrinsic width $\Gamma_0^\downarrow$ was set equal to 5 MeV (full drawn curve) and to 6 MeV (dotted-dashed curve), keeping fixed the rotational frequency at ω = 1.0 MeV. In both calculations displayed in the bottom part of the figure the value $\Gamma_0^\downarrow$ = 5 MeV was used, and ω = 1.0 MeV (full drawn curve) and ω = 1.25 MeV (dashed curve). Note that the FWHM of the corresponding strength functions (shown with dashed lines) is the same. (Adapted from Bracco et al. (1995).)

increases because the "effective deformation" of the system increases,[1] the absolute value of the associated $a_2(E_\gamma)$ increases (results shown with a dashed curve at the bottom row of the Fig. 6.6). The strength function and the $a_2(E_\gamma)$ were calculated with the thermal fluctuation model of shape and orientation in the adiabatic limit (Ormand et al. (1990)).

The Sn-data associated with a restricted range of values of the angular momentum and for a rather narrow temperature interval (T = 1.6–2.0 MeV) are shown in comparison with the statistical model calculations in Fig. 6.7 (left column). The strength function data are displayed on a linear scale.

It is seen that by changing the angular momentum from 40 $\hbar$ to 55 $\hbar$ the width of the giant dipole resonance increases by $\approx$ 2 MeV, that is, one observes

---

[1] The "effective deformation" of the system can change due to an actual change of the shape of the nucleus, or due to a change in the rate with which the nuclear shape and orientation fluctuate.

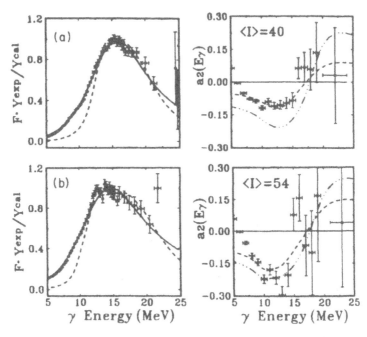

Figure 6.7. Measured strength functions and $a_2(E_\gamma)$ for 2 different angular momenta for ^{109,110}Sn. On the left part of the figure the quantity $F(E_\gamma) * Y_\gamma^{exp}(E_\gamma)/Y_\gamma^{cal}(E_\gamma)$ (see text) is plotted. The full drawn lines correspond to the best fitting single-component Lorentzian functions. In the case of lower angular momenta the obtained width is 10.8 MeV, while at the larger angular momentum values it becomes 12.8 MeV. The dashed lines display results of calculations carried out taking into account adiabatically, thermal fluctuations of the shape and of the orientation of the nucleus (cf. Ch. 10). The results represented by the dotted-dashed lines include only shape fluctuations. (Adapted from A. Bracco et al. (1995).)

a 20% increase of the FWHM, that is by increasing the angular momentum content of the nucleus by $\approx 40\%$. At the same time, the angular distribution (minimum) changes by a factor of 2 (cf. also Fig. 6.7).

Based on the discussion made in connection with Fig. 6.6, one can conclude that the collisional damping width $\Gamma_D^\downarrow$ does not change as a function of angular momentum and that the measured increase of the giant dipole resonance FWHM is mainly caused by the increase of the nuclear deformation as a function of angular momentum.

To make this discussion more quantitative, we compare the data to predictions obtained with a model which takes into account the coupling of the giant dipole resonance to the thermal fluctuations of both the shape and the orientation of the deformed nuclear system (cf., e.g., Gallardo et al. (1985), Ormand et al. (1992)) in the adiabatic limit. The calculations were made for the ensembles of (T,ω) values corresponding to the (E^*,I) regions relevant to the case under discussion. Two sets of calculations were made, one including only shape fluctuations, and the other including both shape and orientation fluctuations (dotted-dashed and dashed curves in Fig. 6.7, respectively). In both cases the intrinsic damping width Γ_0 (cf. caption to Fig. 6.6) was chosen equal to 5 MeV. While the results of the two sets coincide in the case of the strength function, they are different in the case of the $a_2(E_\gamma)$-coefficient. The calculated strength functions, normalized to the data (in arbitrary units) in the centroid region, although slightly narrower than the measured ones correctly display the relative increase with angular momentum as experimentally observed. In contrast, the increase of the magnitude of the $a_2(E_\gamma)$ is not well reproduced by the results of the calculation which includes both shape- and orientation-fluctuations. Furthermore, it is observed that the data associated with the highest values of spin are better reproduced by calculations which only include shape fluctuations. This result suggests that either the fluctuations in orientation are overestimated at high angular momenta or that the collisional damping width has a smaller value than the one used here. This last possibility would lead to a decrease of the GDR width, at variance with the information provided by the measured spectra. It is then likely that, at high rotational frequencies, there is an excess of averaging in the adiabatic model as compared to that required by the experimental findings (cf. Sect. 10.8, in particular 10.8.3).

6.2 Limiting Temperature

Because the giant dipole resonance seems to be rather stable with temperature, it was expected it could be used to probe nuclear structure as a function of temperature, up to the highest values of T (≈ 5–7 MeV) a nucleus can sustain as a self-bound system. A number of experiments were therefore made using projectiles (of mass around 40) with bombarding energies of the order of the Fermi energy per nucleon (i.e., 10 MeV/u to 40 MeV/u). The energy transferred in these experiments to the composite systems, having masses smaller that those corresponding to the complete fusion of the projectile with the target, may be calculated in a spectator-participant model. This model allows to determine the linear momentum transfer fraction as well as the mass and the excitation energy of the composite system from the measured residue velocity.

The first studies of the giant dipole resonance carried out in nuclei with excitation energies larger than 300 MeV, and thus temperature of the order of 4–5 MeV, were made using the reactions ^{40}Ar+^{70}Ge at 15 and 24 MeV/u. The high-energy γ-rays were measured in coincidence with signals from position sensitive parallel plate avalanche counters placed at forward angles. They were used to measure the velocity of the recoiling fusion residues. High-energy γ-ray spectra associated with the most central collisions, namely associated with the largest possible values of the ratio of the recoil velocity v_R to the center of mass velocity v_{CM} were, in the case of the 15 MeV/u reaction, well reproduced both in shape and intensity by a statistical model calculation, assuming that the decaying nuclei were formed in essentially complete fusion reactions with

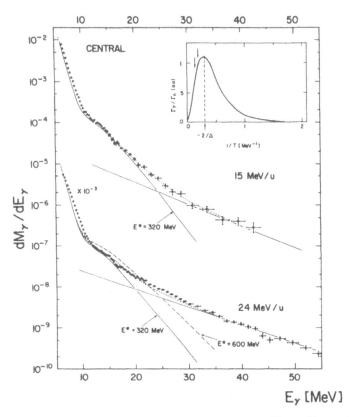

Figure 6.8.   Measured γ-ray spectrum from the heavy ion reaction $^{40}\text{Ar} + {}^{70}\text{Ge} \rightarrow {}^{110}\text{Sn}$ at the bombarding energies 15 and 24 MeV/u. The contributions from pre-equilibrium γ-rays due to the nucleon–nucleon collisions are shown with the straight lines. Gamma-rays from the decay of the giant dipole resonance in the excited nucleus are concentrated in the range $E_\gamma = 10$–25 MeV. The full drawn line shows a spectrum calculated with the statistical model as described in the text, for different excitation energies of the compound nucleus. (Adapted from Gaardhøje et al. (1987).)

excitation energy $E^* = 320$ MeV. By further increase of the excitation energy, as in the case of the 24 MeV/u reaction, the number of γ-rays emitted per fusion reaction was expected to increase. Surprisingly, this was not the case. The yield of γ-rays in the GDR region was found to be the same for the bombarding energies 15 and 24 MeV/u, as can be seen from Fig. 6.8.

Other studies, inspired by these findings and carried out by different groups, have confirmed the results. To be mentioned is the experiment done with the large detection facility MEDEA (cf. Le Faou et al. (1994)) in an effort to determine as precisely as possible the excitation energy of the decaying nucleus through the measurement, not only of the velocities of evaporation residues, but also of the multiplicity and energy spectra of light charged particles. In that case, the reaction $^{36}\text{Ar} + {}^{90}\text{Zr}$ at 27 MeV/u was used, and proton spectra associated with different velocities of the recoiling residues were measured at several angles. The analysis of the proton spectra showed that the apparent temperature of the emitting source of light charged particles increased when larger values of the v_R/v_{CM} ratio were selected. Also in this case, spectra of high-energy γ-rays associated with recoiling residues with the largest values of v_R/v_{CM}, and corresponding to excitation energies $E^* = 350$ and 500 MeV, display a GDR γ-yield which is essentially equal.

The existence of a limiting temperature for the observation of the GDR, considerably lower than the limiting temperature for the nucleus to exist as a bound system, seems to indicate that highly excited nuclei may cool substantially by particle evaporation before the γ-ray emission from the GDR can occur. This result can be understood in terms of the fact that it only takes few 2 fm/v_F $\approx 2 \times 10^{-23}$ sec to thermalize the single-particle motion after the fusion event has taken place (i.e., few collisions per nucleon, the average nucleon distance being 2 fm), while it takes $\hbar/\Gamma_D^{\downarrow} \approx 1.3 \times 10^{-22}$ sec for the giant dipole resonance to become a thermal excitation of the compound nucleus. The system has thus ample time to cool down by particle emission, before the GDR can decay.[2] The temperature at which the decay width of the compound nucleus for particle emission $\Gamma_{CN}^{\uparrow}$ becomes of the order of $\Gamma_D^{\downarrow}$, defines a limiting temperature for γ-decay of the GDR (cf. Bortignon et al. (1991)).

This phenomenon can be better understood assuming, for simplicity, that in a hot nucleus there are only two classes of states: a) the compound nucleus states, b) the compound nucleus states with a GDR build on top of it (cf. Figs. 1.3 and 6.9). Consequently, and in keeping with the labelling of Fig. 6.9, to each state C there corresponds a state D shifted up by the energy of the GDR. The transition rates between these two classes of states are denoted λ and μ respectively, and the ratio λ/μ, because of the principle of detailed balance, is equal to the ratio of the associated level densities

$$\frac{\lambda}{\mu} = \frac{\rho_D}{\rho_C} = \frac{\rho_C(E^* - E_{GDR})}{\rho_C(E^*)} \ll 1. \tag{6.6}$$

Both classes of states can decay by particle emission and only the dipole states by γ-emission. The time dependent probabilities for the system to be in either class of states are denoted by P_C and P_D and satisfy the equations

$$\frac{dP_D}{dt} = -(\mu + \gamma_\gamma + \gamma_{ev})P_D + \lambda P_C, \tag{6.7}$$

---

[2] The same line of reasoning applies to the restoration of isospin symmetry in compound nuclei at excitation energies of the order of 50–100 MeV, as pointed out by Wilkinson (1956) and experimentally tested in measurements of the suppression of the GDR decay in isospin zero compound nuclei (Behr et al. (1993), and Snover (1993)). This is because the compound nucleus does not live long enough for the isospin symmetry to be broken by the weak Coulomb interaction.

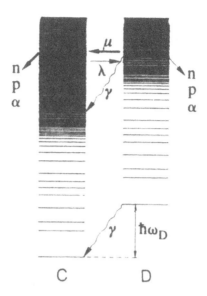

Figure 6.9. Schematic representation of nuclear energy levels. Levels C are ordinary compound nucleus states. States D are giant dipole resonance states built on C states, and therefore shifted by the giant dipole resonance energy. Each type of state can decay by particle emission. Only D states can decay by γ emission. Transition rates between C and D states are denoted by λ and μ. (Adapted from Bortignon et al. (1991).)

$$\frac{dP_C}{dt} = -(\lambda + \gamma_{ev})P_C + \mu P_D. \tag{6.8}$$

In the above expressions, γ_γ and γ_{ev} are the decay rates for γ-rays and particles, respectively. Because of the difference in level density, $\lambda \ll \mu$. The above equations can be solved assuming that the system starts in a state C, so that the γ-emission probability integrated over the resonance is given by

$$P_\gamma = \frac{\gamma_\gamma}{\gamma_{ev}}\left(\frac{\lambda}{\gamma_{ev} + \lambda + \mu}\right) \approx \frac{\gamma_\gamma}{\gamma_{ev}}\left(\frac{\lambda}{\gamma_{ev} + \mu}\right). \tag{6.9}$$

Below the critical excitation energy E_{crit}, $\gamma_{ev} \ll \mu$,

$$P_\gamma(E^* < E_{crit}) = \frac{\gamma_\gamma}{\gamma_{ev}}\frac{\lambda}{\mu}, \tag{6.10}$$

and well above this excitation energy, $\gamma_{ev} \gg \mu$

$$P_\gamma(E^* > E_{crit}) = \frac{\gamma_\gamma \lambda}{\gamma_{ev}^2} \ll P_\gamma(E^* < E_{crit}). \tag{6.11}$$

By equating the decay width associated with the decay of the dipole states into the compound states, to the spreading width of the giant dipole resonance, a limiting excitation energy for the observation of the GDR can be calculated. Evaluating the temperature dependence of $\Gamma_{CN}^\uparrow$ (cf., e.g., Pülhofer (1977)) and making use of the fact that $\Gamma_D^\downarrow$ is independent of temperature (cf. Chs. 9 and

10 and also the previous section) one obtains, in the case of ^{110}Sn, $E^*_{lim} \approx 250$ MeV, in overall agreement with the experimental findings.

The main limitation of the model lies in the assumption that at $t = 0$, $P_D = 0$. In fact, dipole modes can, in principle be present in the precompound stages of the reaction, due to isospin asymmetry between target and projectile (cf., e.g., Chomaz et al. (1993)). However, the corresponding dipole modes have an energy considerably lower than that associated with the compound nucleus GDR because, at low impact parameter for which v_R/v_{CM} display the largest values, the vibration is mainly along the beam axis, over a length of the order of the sum of the radii (Bortignon et al. (1995) and refs. therein).

6.3 Fissioning Nuclei

Measurements of neutron multiplicities before and after nuclear scission seem to indicate that fission is a cold process, significantly slower than expected on the basis of the statistical model of nuclear decay (Hilscher and Rossner (1992)), occurring late in the decay of the excited compound nuclei. The same indication was provided by charged particle measurements. Similar information has been obtained (Thoennessen et al. (1987)) from the comparison of the multiplicity associated with γ-decay of the giant dipole resonance built on fissioning compound nuclei and statistical model predictions. In fact, it has been observed that the γ-decay of the GDR competes favorably with the fission process.

If one were to explain the observations in terms of GDR parameters, one would need to assume that the associated oscillator strength largely exceeds the classical sum rule value, an ansatz which cannot be justified within the present theoretical understanding of the GDR. Conversely, as for the neutron and charged-particle pre-fission emission, it is possible to explain the γ-multiplicity data in terms of a large viscosity slowing down the fission process by more than an order of magnitude as compared with the statistical model prediction (Bohr–Wheeler fission rate) at temperatures of the order of 1–2 MeV.

In the case of neutron emission it is possible to separate the neutron yield associated to pre- and post-fission phenomena by appropriately choosing, the angles at which the particles are measured. On the other hand, the energy spectrum of high-energy γ-rays emitted from fissioning nuclei contains both contributions from pre- and post-fission process. The spectrum associated with the gamma-emission prior to fission reflects the high-temperature features of the nucleus formed in the fusion reaction and contains essentially no low-energy gamma rays. Its spectral shape is characterized by the GDR with centroid energy $\approx 80A^{-1/3}$ (cf. Eq. (2.1)), that is, lower than that of the fission fragments $\approx 80(A/2)^{-1/3}$. Making use of the accumulated systematics on the GDR properties at finite temperature and angular momentum and of the statistical model it is possible to obtain information on the prefission process and, in principle, on prefission nuclear deformations. This information is not accessible to standard spectroscopy studies or to particle-decay studies.

6.4 The Pre-Fission γ-Decay

A study of high-energy γ-decay from a very fissile hot nucleus was conducted for ^{224}Th formed in the fusion reaction ^{16}O + ^{208}Pb at the three different bombarding energies of 100, 120 and 140 MeV. The corresponding excitation energy and average angular momentum of the compound nucleus lie within the intervals of 44 to 82 MeV and of 25 to 43 $\hbar$, respectively (Butsch et al. (1991)).

The measured high-energy γ-ray spectra were analyzed with the statistical model. In this type of analysis two components have to be considered, one associated with the pre-fission and the other with the post-fission phenomenon. The component corresponding to post-fission emission contains two contributions. The first is due essentially to low-energy γ-rays emitted by the fission fragments after particle evaporation has lowered their intrinsic excitation energy, below particle threshold.

The excitation energy of each fragment E_{fis}^* is given by

$$E_{fis}^* = \frac{1}{2}(E_{CN}^* + Q_{fis} - E_{kin}) - E_{rot} - E_{def}, \tag{6.12}$$

where E_{CN}^* is the excitation energy of the compound nucleus, Q_{fis} is the Q-value for the fission process assuming a particular initial mass division (e.g., symmetric fission). The quantity E_{kin} is the total kinetic energy of the fission fragments ($E_{kin} = E_{kin}^1 + E_{kin}^2$), a quantity which does not depend on the excitation energy of the compound nucleus but is fixed by the Coulomb repulsion of the two fragments after scission. The quantities E_{def} and E_{rot} are the energies associated with the deformation and the rotation of the fission fragments.

The computation of the pre-fission part of the high-energy γ-spectrum is based on the fission rate as determined from phase space considerations and neglects any dynamical effect. The corresponding fission rate is given by the Bohr and Wheeler (Bohr and Wheeler (1939)) width

$$\Gamma_{fiss}^{BW} = \frac{1}{\rho(E_i, I_i)} \int_0^{E_i - E_B} \rho(E_i - E_B - E_{kin})dE_{kin}, \tag{6.13}$$

where E_B is the fission barrier, E_i is the excitation energy of the compound nucleus ($E_i = E_{CN} - E_{CN}^{rot}$, namely the part of the excitation energy of the compound nucleus that does not go into rotational energy). The compound nucleus level density is denoted by ρ.

The calculations made following this approach were able to reproduce both the low- and high-energy part of the γ-spectrum, but could not explain the observed gamma-ray strength in the energy region around 11–12 MeV. This energy corresponds to that of the GDR built on the heavy fissile compound nucleus. As in the case of neutron emission, to describe the measured spectra the fission probability in the first steps of the compound decay had to be reduced in order to enhance the contribution of the pre-fission emission in the total gamma-ray spectrum. This reduction factor was found to be dependent on the excitation energy of the compound nucleus.

An experimental confirmation of the strong pre-fission contribution of the GDR to the γ-spectrum was obtained through measurements of the γ-ray angular anisotropy defined as a ratio of the high-energy γ−rays measured at $0°$ and $90°$, with respect to the direction of the angular momentum of the compound nucleus, as deduced from the directions of the fission fragments, detected in coincidences with the high-energy γ-rays. In the case of the experiment of Fig. 6.10 the kinetic energy of fission fragments was measured in two pairs of Si detectors located in the plane perpendicular to the beam direction. The coincidence between the high-energy γ-detector and the fission fragment detectors aligned with it, defines the angular correlation of the gamma rays at $90°$, while the other combination does it for $0°$. The observed anisotropies at $E_\gamma = 10$–11 MeV are consistent with those expected for the low-energy component of the GDR in a heavy deformed nucleus.

As in the case of particle emission, an explanation of the measured yield of the γ-decay of the giant dipole resonance in fissioning nuclei can be obtained assuming that nuclear fission is a strongly dissipative process (see P. Paul and M. Thoennessen (1994) for a review). The dissipation phenomenon controls the fission time scales and can be described making use of Fokker–Plank equations (Grangè and Weidenmüller (1980)) describing the time evolution of the collective coordinates during the fission process. A scale for the dissipation is provided by the nuclear friction coefficient $f = \alpha/2\omega$. In this expression α is the reduced dissipation coefficient in units of 10^{-21} sec while ω (in the same units) is the knocking rate frequency of the fissioning system moving in the average potential inside the saddle point. Critical, overdamped and underdamped situations correspond to $f = 1$, $f > 1$ and $f < 1$, respectively. The presence of dissipation influences the fission process as well as particle- and γ-ray decay rates in the three different stages of the fission process. These stages are: a) the statistically equilibrated compound nucleus from the time of its formation to that in which it reaches the saddle point b) the compound nucleus at the saddle, c) the compound nucleus from the saddle to scission. Inside the saddle, the fission flux builds up with a time constant τ_D fixed by the friction coefficient f. As the fission motion reaches the saddle, the viscous diffusion process leads to a fission width which is reduced, relative to the non-viscous width, as originally proposed by Kramers (1940). Thus, the fission

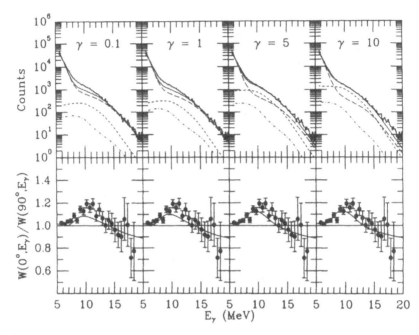

Figure 6.10. Comparison of statistical model calculations with different dissipation coefficient γ ($\equiv f$, cf. text) for the measured γ-ray spectra and angular distribution for the reaction $^{16}O + ^{208}Pb$ at incident energy of 140 MeV. The total γ-ray spectrum (solid) in the top panel consists of γ-ray contributions from pre-scission (short-dashed), saddle-to-scission (dot-dashed), and post-scission (long-dashed). (Adapted from Butsch et al. (1991).)

width at the saddle is given by,

$$\Gamma_{fis}(t) = \Gamma_{fiss}^{BW}(\sqrt{1+f^2} - f)[1 - \exp(t/\tau_D)]. \qquad (6.14)$$

For a value of $f = 2$, reductions of the order of 80% are obtained. The motion along the path from saddle to scission is also expected to be viscous, the associated friction coefficient likely displaying a value different than that before the saddle.

The dissipation effects in the fission process have been studied first and most intensively in thorium. The gamma-ray spectrum of ^{224}Th at $E^* = 90$ MeV obtained with the reaction $^{16}O + ^{208}Pb$ at a bombarding energy of 140 MeV, was analyzed varying the friction coefficient. The sensitivity of the spectral fits to the dissipation can be appreciated from Fig. 6.10. Here, the same data set is compared to calculations with identical parameters for the giant dipole resonance and for the level density, but with friction coefficients increasing from a very underdamped situation ($f = 0.1$) through critically damped ($f = 1$) to a strongly overdamped situation ($f = 10$). The best fit to the data was obtained with a value of $f = 5$–10. Similar results were obtained from the analysis of neutron multiplicity measurements (Hilscher and Rossner (1992)). The strong dissipation slows down the fission process significantly from 4.2 $\times$

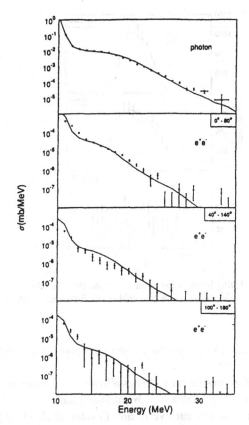

Figure 6.11. Measured γ-ray yield (top) and dilepton yields (bottom three panels) for the T = 0, ^{16}O + ^{12}C reaction. The dilepton yields are shown for three different opening angle between the two leptons. The full lines represent the cross sections calculated using an extended version of the statistical model code CASCADE folded with the measured response function of the spectrometers used. (Adapted from Buda et al. (1995).)

10^{-20} sec ($f = 0$) to 6.4×10^{-19} sec (averaged over the various decay steps). In keeping with these results, the fission process takes place after the nucleus has cooled down, mainly through particle- but also by γ-emission.

The excess of pre-fission GDR γ-rays in ^{224}Th originates overwhelmingly from inside the barrier, as can be seen from Fig. 6.10. On the other hand, in the case of heavier nuclei, the contribution to the γ-yield emitted from saddle to scission becomes progressively dominating in the pre-fission emission. The study of the high-energy γ-rays emitted by the fissioning compound nucleus ^{240}Cf formed at an excitation energy $E_x = 93$ MeV, using the reaction ^{32}S + ^{208}Pb, has showed that the value of the dissipation parameter associated with the saddle-to-scission is 4.3, and thus smaller than that found in stage a) (equilibration to saddle), in which case $f = 10$. From this result, it was

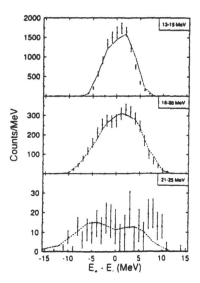

Figure 6.12. Energy sharing between the positron and the electron. The dashed lines represent the internal conversion of the measured photon yield in the Born approximation. (Adapted from Buda et al. (1995).)

deduced that the time it takes for the system to evolve from saddle to scission is $\tau_{ssc} = 26 \times 10^{-21}$.

The enhanced GDR pre-fission yield has allowed to investigate the properties of heavy nuclei with large values of the atomic number Z at high excitation energies, nuclei that do not survive fission (Tveter et al. (1996)). From the observation of the pre-fission GDR multiplicity, it was concluded that the lifetime of the composite system is of the order of 10^{-20} to 10^{-19} sec.

6.5   Search for the Giant Monopole Resonance in Hot Nuclei

The study of the isoscalar giant monopole resonance in hot nuclei is of particular interest because the energy of this state can be related to the compressibility of the nucleus and, by a suitable extrapolation, to the compressibility of nuclear matter (cf. Sect. 2.3). This resonance is particularly difficult to observe since the single photon decay channel is not available. In fact, monopole transitions proceed practically only by electron–positron (e^+, e^-) pair emission. The branching ratio for the monopole pair transitions from the GMR induced in a typical heavy-ion reaction is of the order of 10^{-8}. The main difficulty in studying the monopole transitions is, however, connected not with their low yields but with the fact that all the higher-multipolarity-transitions may also proceed via (e^+, e^-)-pair emission.

Furthermore, a dilepton spectrum from a hot nucleus is expected to be dominated by the pair decay of the GDR which is ≈ 3 to 4 orders of magnitude stronger than the corresponding E0 pair decay from the monopole giant resonance.

A search for the isoscalar monopole strength has been carried out by Buda et al. (1995) for the ^{28}Si compound nucleus, formed at $E_x = 50$ MeV in the ^{16}O$+^{12}$C compound nucleus reaction at beam energy of 82 MeV. This process is particularly suited to reduce the contribution of the giant dipole resonance to the dilepton decay process. This is because it produces a self-conjugated compound nucleus (with N = Z), populated by isospin 0 entrance channels. Because of the small isospin mixing present in compound nuclear states (cf. Sect. 6.2), the yield of high-energy γ-rays from the isovector giant dipole resonance was found for such reactions to be strongly inhibited as compared with compound nuclei with N $\neq$ Z populated by entrance channels with isospin different than zero (Harakeh et al. (1986)).

A dilepton magnetic spectrometer, having detection capability between 10 and 40 MeV, was constructed for this purpose and used in that measurement. It allows to distinguish the electrons from the positrons and the two leptons can be detected with opening angles ranging from 0° to 180°.

The results obtained are shown in Figs. 6.11 and 6.12. The first figure shows the measured photon cross section in comparison with that of the $e^+ e^-$ pair at three different opening angles. The data are compared to the calculated internal pair conversion process of the photon decay channel in the Born approximation. This comparison shows that the measured dilepton strength stems from the internal pair conversion of the photon decay from the GDR and GQR decays. The same conclusion is obtained comparing calculated and measured energy sharing between the positron and the electron for different transition energy ranges (cf. Fig. 6.12). Further improved, highly selective techniques for observing the monopole pair transitions are needed in order to identify the GMR in hot nuclei from dilepton spectra.

7

Concepts of Statistical Physics

A number of powerful methods exist in statistical mechanics to describe the interweaving of the variety of elementary modes of excitation taking place in a quantal many-body system. Among them, one can mention the Static Path approximation (cf., e.g., Feynmann and Hibbs (1965)), the Monte-Carlo Shell Model Method (Koonin et al. (1997)), and the thermal Green-function formalism (cf., e.g., Mahan (1981)). The first two approaches have not yet been systematically used to describe nuclear motion at finite temperature. We shall briefly discuss them in the present chapter. The Matsubara techniques, which have been applied to the study of a large variety of phenomena, are discussed in Ch. 9.

7.1 The Variety of Ensembles

Let us assume that a highly excited nucleus is formed at an excitation energy E^*, characterized by a set of quantum numbers like angular momentum I, parity π, etc., where all the degrees of freedom of the system are in equilibrium (compound nucleus). Furthermore, let us assume an equal probability distribution of all the states with these quantum numbers. The probability p_i to find the compound nucleus in a state i at an energy E_i, angular momentum I_i and parity π_i can then be written as

$$p_i = \frac{\delta(E_i - E^*)\delta(I_i - I)\delta(\pi_i - \pi)}{\rho(E^*)}, \tag{7.1}$$

where $\rho(E^*)$ is the density of states. The conservation of other quantum numbers introduces other multiplicative δ functions. The resulting probability distribution is called microcanonical.

It is more convenient to use exponential forms of the probability distributions rather than δ–functions. This can be accomplished in keeping with the fact that the nuclear level density $\rho(E^*)$ can be written in the form

$$\rho(E^*) = \rho_o \exp\left(S(E^*)\right), \tag{7.2}$$

where $S(E^*)$ is the entropy. Consequently the quantity introduced in Eq. (5.3), can now be written as

$$T = \left(\frac{dS}{dE^*}\right)^{-1},\qquad(7.3)$$

The next step in our reasoning is to recognize that the function

$$\rho(E^*)\exp\left(-\beta E^*\right),$$

depending on the parameter β, is sharply peaked at the value E^* determined by the value of β given by

$$\beta = \frac{dS}{dE^*} = T^{-1}.$$

Therefore, the microcanonical distribution defined in Eq. (7.1) may be conveniently replaced by the distribution

$$\frac{\rho(E^*)\exp\left(-\beta E^*\right)}{Z},\qquad(7.4)$$

called the canonical distribution. The quantity $Z(T)$ is the partition function

$$Z(T) = \int_0^\infty \rho(E^*)\exp(-\beta E^*)dE^* = \sum_i \exp\left(-\beta E_i\right).\qquad(7.5)$$

In calculating the thermal average of an observable $\hat{O}$, one needs to take the trace over all the states of the system, i.e.,

$$\begin{aligned}\langle\hat{O}\rangle_T &= \frac{1}{Z}\sum_i \langle i|O|i\rangle \exp\left(-\beta E_i\right)\\ &= \frac{1}{Z}\int_0^\infty O(E^*)\rho(E^*)\exp(-\beta E^*)dE^*.\end{aligned}\qquad(7.6)$$

This integral can be calculated in the saddle-point approximation, the saddle-point E^* being given by $\beta = d\ln\rho(E^*)/dE^* = T^{-1}$, that is, the value at which the canonical distribution defined in Eq. (7.4) is sharply peaked. The relation between the microcanonical and the canonical distribution ensues.

The grand canonical probability distribution is obtained relaxing particle-number conservation in the canonical distribution. Eventually, the density operator

$$\hat{D} = \frac{\exp\left(-\beta(\hat{H} - \mu\hat{N})\right)}{Z},\qquad(7.7)$$

is introduced, $\hat{H}$ being the Hamiltonian describing the system and $\hat{N}$ the number of particle operator. The temperature β^{-1} and the chemical potential μ are fixed, while energy and number of particle are not. Only the corresponding average values are defined. The partition function Z is now the trace of the numerator in Eq. (7.7), that is,

$$Z = Tr\exp\left(-\beta(\hat{H} - \mu\hat{N})\right)\qquad(7.8)$$

In the above discussion the basic ansatz is that of equilibration, and not the existence of a heat bath (cf. e.g., Levit (1988)).

In heavy ion fusion reactions leading to highly excited nuclei, a sizeable amount of angular momentum I can be transferred from the relative to the intrinsic motion with a continuous distribution. To take into account angular momentum conservation, one can define the partition function as

$$Z(T, I) \equiv Tr \exp\left(-\beta F(\beta, I)\right) = Tr P_I \exp\left(-\beta(\hat{H} - \mu\hat{N})\right), \qquad (7.9)$$

where P_I is the angular momentum projection operator. Introducing the angular velocity $\vec{\omega}$, one finds

$$F(T, \vec{\omega}) = -T \ln Tr \exp\left(-\beta(\hat{H} - \mu\hat{N} - \vec{\omega} \cdot \vec{I})\right). \qquad (7.10)$$

In deriving this expression use is made of the relation

$$\begin{aligned} \rho(E^*, I) &= \rho(E^*, M = I) - \rho(E^*, M = I + 1) \\ &\approx (\partial/\partial M)\rho(E, M)|_{M = I + 1/2}, \end{aligned} \qquad (7.11)$$

to describe the nuclear density of states at given angular momentum and projection I, M. In other words, in the free energy of a rotating nucleus referred to the fixed-body frame of reference, the Hamiltonian is replaced by the Routhian.

7.2 Level Density and Partition Function

In the previous section we have essentially made use only of the fact that the level density is an exponentially growing function of the excitation energy. In this section we shall discuss in more detail the variety of properties of this function.

Experimentally, the level density is "measured" by counting levels. For example, the sharp resonances in slow neutron capture reactions or the states excited in (p,p') inelastic scattering, up to particle separation energies. At higher excitation energies, one must rely on the expression obtained in a model, and fit one or more parameters to reproduce, e.g., the particle evaporation spectra (cf. Sect. 2.5). At still higher excitation energies, and because of the very many degrees of freedom involved, the Fermi gas model is expected to be useful.

7.2.1 The Fermi gas model

In the Fermi gas model, the level density is obtained by counting the number of ways in which the excitation energy E^* of a system with A-particle can be distributed among single-particle states described in terms of plane-wave wavefunctions. For $N \approx Z \approx A/2$, the result is (Bohr and Mottelson (1969)),

$$\rho(A, E^*) = \frac{6^{1/4} g_0}{12(g_0 E^*)^{5/4}} \exp(2\sqrt{aE^*}). \qquad (7.12)$$

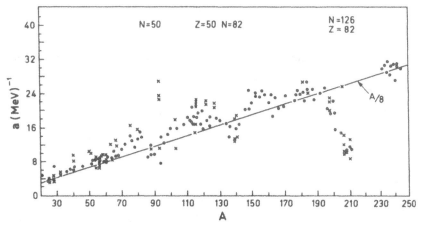

Figure 7.1. The level density parameter a as a function of the mass number A. It has been determined by comparison with the average spacing observed in slow neutron resonances (open circles) and from evaporation spectra (crosses). (From Bohr and Mottelson (1969) and Erba et al. (1961).)

The level density parameter a is defined as

$$a = \frac{\pi^2}{6} g_0, \qquad (7.13)$$

in terms of the single-particle level density g_0 at the Fermi energy ϵ_F, in keeping with the fact that, for a fermion system, temperature can lead to modifications only to the Fermi surface. The quantity g_0 has the value

$$g_0 = \frac{3}{2} \frac{A}{\epsilon_F}, \qquad (7.14)$$

and is therefore proportional to the nucleon effective mass m^* (cf. Eq. (4.23)) at the Fermi energy ($\epsilon_F = \hbar^2 k^2 / 2m^*$). Using the bare mass value and the Fermi momentum $k_F \approx 1.37$ fm^{-1} one obtains

$$a \approx \frac{A}{16} \text{ MeV}^{-1}. \qquad (7.15)$$

The empirical value, obtained for an excitation energy of the order of 10 MeV, is, as shown in Fig. 7.1, a factor two larger, namely

$$a \approx \frac{A}{8} \text{ MeV}^{-1}, \qquad (7.16)$$

with considerably smaller values for closed-shell nuclei.

The value of a given in Eq. (7.16) is found to decrease as a function of the excitation energy (cf. Gonin et al. (1989) and refs. therein), and to achieve the value $a \approx \frac{A}{12}$ MeV^{-1} for excitation energies larger than 1 MeV per nucleon (cf. Fig. 7.2).

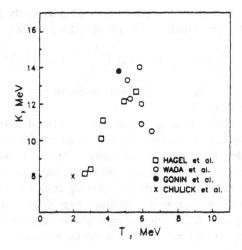

Figure 7.2. Empirical values of the parameter $K = A/a$ as a function of the temperature. (Adapted from Gonin et al. (1989).)

Using Eqs. (7.12) and (7.16) one obtains, for a nucleus with $A \approx 100$ at an excitation energy $E^* \approx 50$ MeV, a level density $\rho \approx 10^{20}$ MeV^{-1}. This value is very large as compared to the energy resolution with which one can specify the state of the system. Consequently, the definition of nuclear temperature introduced in Eq. (7.3) is well defined.

From Eqs. (7.3) and (7.12) one can write

$$T^{-1} = -\frac{5}{4}\left(E^*\right)^{-1} + \left(\frac{a}{E^*}\right)^{1/2}. \qquad (7.17)$$

For excitation energies greater than the energy of the first excited state of the system ($g_0 E^* \gg 1$), the second term dominates leading to

$$E^* = aT^2. \qquad (7.18)$$

The main results contained in Eqs. (7.12) and (7.18) can be qualitatively derived (Bohr and Mottelson (1969)) reasoning in terms of the average occupation factors

$$n_F(\epsilon) = \frac{1}{1 + \exp\left(\frac{\epsilon - \epsilon_F}{T}\right)}, \qquad (7.19)$$

in a Fermi gas at a temperature T, associated with a single-particle state of energy ϵ. Heating the cold system to a temperature T, changes the average occupation numbers of the single-particle states in an energy region of order T around ϵ_F, exciting the average number of quasiparticles

$$v \approx g_0 T.$$

Each quasiparticle contributes an average excitation energy of order T, leading to the result $E^* \approx g_0 T^2$, as in Eq. (7.18). Furthermore, the main term in the level density will be given by the number of ways in which the v quasiparticles can be distributed over a number of levels of order $2v$, that is

$$\rho \sim \frac{(2v)!}{[(v)!]^2} \sim \exp(v) \sim \exp\left(\frac{E^*}{T}\right) \sim \exp(aE)^{1/2}. \qquad (7.20)$$

This function displays the same exponential factor of the density function obtained in Eq. (7.12).

In the previous considerations, the nucleus has been described in terms of independent motion of the nucleons moving in an infinite system subject to the Pauli principle. Consequently, all many-body effects but those associated with fermion statistics have been neglected, in particular, those associated with the presence of a nuclear surface displaying collective vibrations. Consequently, the effects of both small-amplitude quantal fluctuations as well as large-amplitude thermal fluctuations, of particular interest within the context of the present monograph, are not included in the expression of the level density given by Eq. (7.12).

A powerful method to take these effects into account has been developed making use of functional-integral techniques (cf., e.g., Alhassid and Zingman (1984), P. Arve et al. (1988), Puddu et al. (1991) and refs. therein). In what follows, we shall discuss this method and provide examples of its applications.

7.2.2 Role of correlations

The level density for a system with A particles at excitation energy E^* is given by

$$\rho(E^*, A) = \sum_{n,i} \delta(A - n)\delta(E^* - E_i(n)), \qquad (7.21)$$

where $E_i(n)$ is the energy of the ith state of the n-particle system. In terms of the partition function Z calculated in the grand canonical ensemble (cf. Eq. (7.8))

$$Z(\alpha, \beta) = Tr[\exp{(-\beta\hat{H} + \alpha\hat{N}))}],$$

the level density can be written as (Bohr and Mottelson (1969))

$$\rho(E^*, A) = \frac{1}{2\pi D^{1/2}} \exp{(\beta_o E^* - \alpha_o A + \ln Z(\beta_o, \alpha_o))}, \qquad (7.22)$$

where β_o and α_o are fixed by the average excitation energy and the average particle number via the conditions

$$E^* = -\frac{\partial \ln Z(\alpha, \beta)}{\partial \beta}|_{\beta_o}. \qquad (7.23)$$

$$A = \frac{\partial \ln Z(\alpha, \beta)}{\partial \alpha}|_{\alpha_o}. \qquad (7.24)$$

The quantity D is the determinant of the matrix

$$\begin{pmatrix} \frac{\partial^2 \ln Z(\alpha,\beta)}{\partial \beta^2} & \frac{\partial \ln Z(\alpha,\beta)}{\partial \beta}\frac{\partial \ln Z(\alpha,\beta)}{\partial \alpha} \\ \frac{\partial \ln Z(\alpha,\beta)}{\partial \alpha}\frac{\partial \ln Z(\alpha,\beta)}{\partial \beta} & \frac{\partial^2 \ln Z(\alpha,\beta)}{\partial \alpha^2} \end{pmatrix},$$

calculated for β and α equal to β_o and α_o respectively. Different approximations for the level density are produced by different approximations for the partition function.

In a non-interacting fermion system, one obtains, for the partition function

$$Z(\alpha, \beta) = \prod_k (1 + \exp(\alpha - \beta\epsilon_k)), \qquad (7.25)$$

where ϵ_k is the energy of the single-particle state k. Starting from this expression, introducing the smooth average density of one-particle states $g(\epsilon)$ (with $g_0 = g(\epsilon_F)$) one finds, by substitution in Eq. (7.22), the Fermi gas expression of the level density given in Eq. (7.12).

Functional-integral techniques have been used, to include the effects of correlation and fluctuations associated with both small-amplitude quantal fluctuations as well as with large-amplitude thermal fluctuations, in the calculation of the partition function. In what follows, we shall briefly review the main results of this approach.

Assuming a nuclear Hamiltonian $\hat{H}$ which contains, at most, two-body terms, it is always possible to find a set of one-body operators $\hat{Q}_\nu$, such that

$$\hat{H} = \sum_\nu \epsilon_\nu - \frac{1}{2}\sum_\nu \chi_\nu \hat{Q}_\nu^2. \qquad (7.26)$$

To simplify the notation, we will often consider a Hamiltonian of the type

$$\hat{H} = \hat{H}_o - \frac{1}{2}\chi\hat{Q}^2. \qquad (7.27)$$

Furthermore, the operator $\hat{U} = \exp(-\beta\hat{H})$ appearing in the expression for the partition function, Eq. (7.8), is similar in structure, to a time propagator in quantum mechanics, but for an imaginary time defined by $\beta = it$. It is convenient to write the operator $\hat{U}$, as the exponential of a one-body operator. In this case, the expectation value of $\hat{U}$ in a trial wave function of determinant type, gives rise to another determinant (Thouless (1962)). Trace of determinants are simple to calculate making use of the Hubbard–Stratonovich transformation based on the identity

$$\exp\left[\frac{1}{2}\lambda\hat{O}^2\right] = \sqrt{\frac{|\lambda|}{2\pi}} \int d\sigma \exp\left[-\frac{1}{2}\lambda\sigma^2 + s\sigma\lambda\hat{O}\right], \qquad (7.28)$$

where $s = \pm 1$ if $\lambda > 0$ and $s = \pm i$ if $\lambda < 0$, and where σ is an auxiliary field (Hubbard (1959), Stratonovich (1957)). One can then write

$$\hat{U} = \exp(-\beta \hat{H}) \approx \int \mathcal{D}(\sigma) \exp\left[-\frac{1}{2}\beta \sum_\nu |\chi_\nu|\sigma_\nu^2\right] \exp[-\beta \hat{h}(\sigma)], \quad (7.29)$$

with

$$\mathcal{D}(\sigma) = \prod_\nu \sqrt{\frac{\beta|\chi_\nu|}{2\pi}} d\sigma_\nu,$$

and

$$\hat{h}(\sigma) = \sum_\nu (\epsilon_\nu + s_\nu \chi_\nu \sigma_\nu)\hat{Q}_\nu.$$

The quantity σ_ν is the auxiliary field associated with the one-body operator $\hat{Q}_\nu$. In Eq. (7.29) we used the sign $\approx$, because, although Eq. (7.28) is an identity, when we apply it to rewrite the operator $\hat{U}$ as in Eq. (7.29), we disregard the commutators between kinetic and potential energy terms, an approximation which becomes progressively less severe as the excitation energy increases.

In this approach, the partition function for the Hamiltonian introduced in Eq. (7.26) can be written as a functional integral (Kerman et al. (1981) and (1983), Negele (1982))

$$\begin{aligned} Z &= \int \mathcal{D}[\sigma(\tau)] \exp\left(-\frac{\chi}{2} \int_0^\beta d\tau \sigma^2(\tau)\right) \times \\ &\quad Tr \left[\mathcal{T} \exp\left(-\int_0^\beta d\tau[\hat{H}_0 - \chi\sigma(\tau)\hat{Q}] + \alpha\hat{N}\right)\right], \quad (7.30) \end{aligned}$$

over the auxiliary field $\sigma(\tau)$, where $t = i\tau$ plays the role of an imaginary time, and $\mathcal{T}$ is the time-ordering operator. Different approximations may be used in the evaluation of this integral. The lowest level of approximation corresponds to consider a non-interacting system, that is to set $\chi = 0$. In this case, the result given in Eq. (7.25) is obtained.

The next level of approximation to calculate the integral in Eq. (7.30) is the so-called static path approximation (SPA), where large-amplitude thermal fluctuations are included. In this case, one replaces the field $\sigma(\tau)$ by its time average

$$\bar{\sigma} = \frac{1}{\beta} \int_0^\beta d\tau \sigma(\tau),$$

and neglects the contribution proportional to $[\sigma(\tau) - \bar{\sigma}]$ to the trace in Eq. (7.30). One thus obtains

$$Z = \left(\frac{\beta\chi}{2\pi}\right)^{1/2} \int d\bar{\sigma} \exp(-\frac{\chi}{2}\beta\bar{\sigma}^2$$

$$+ \sum_{\nu} \ln[1 + \exp(-\beta\epsilon_\nu(\bar{\sigma}) + \alpha)]). \tag{7.31}$$

The quantities ϵ_ν are the energies of the single-particle states ν calculated by diagonalizing, for every value of $\bar{\sigma}$, the single-particle Hamiltonian $\hat{h} = \hat{H}_0 - \chi\bar{\sigma}\hat{Q}$. Calculating the integral over the field $\bar{\sigma}$ in the saddle-point approximation, the mean-field expression for the partition function is obtained

$$Z = \exp\left(-\frac{\chi}{2}\beta\sigma_o^2 + \sum_{\nu}\ln[1 + \exp(-\beta\epsilon_\nu(\sigma_o) + \alpha)]\right), \tag{7.32}$$

with $\sigma_o = \sum_\nu n_\nu Q_{\nu\nu}$. The quantities n_ν are the occupation factors of the single-particle states ν, and $Q_{\nu\nu}$ the diagonal matrix elements of the operator $\hat{Q}$. As compared to the partition function for the non-interacting case (Eq. (7.25)), the contribution of the mean field σ_o is noted.

Quantal, small-amplitude fluctuations, can be considered by expanding the free energy associated with the trace in Eq. (7.30) around $\bar{\sigma}$, up to second order in the difference $[\sigma(\tau) - \bar{\sigma}]$. In this way, the time dependence of the field $\sigma(\tau)$ is taken into account, as seen by its Fourier expansion

$$\sigma(\tau) = \bar{\sigma} + \sum_{m\neq 0} \eta_m \exp(-i\frac{2\pi}{\beta}m\tau).$$

The Gaussian integral over the variables η_m can be performed to give

$$\begin{aligned}
Z &= \left(\frac{\beta\chi}{2\pi}\right)^{1/2} \int d\bar{\sigma} \exp(-\frac{\chi}{2}\beta\bar{\sigma}^2 \\
&\quad + \sum_{\nu}\ln[1 + \exp(-\beta\epsilon_\nu + \alpha)])\mathcal{C}(\bar{\sigma}),
\end{aligned} \tag{7.33}$$

where

$$\mathcal{C}(\bar{\sigma}) = \prod_{m>0}\left(1 - \chi\sum_{\mu\nu}\frac{|Q_{\mu\nu}|^2(n_\mu - n_\nu)(\epsilon_\nu - \epsilon_\mu)}{(\epsilon_\nu - \epsilon_\mu)^2 + (2\pi m/\beta)^2}\right)^{-1}. \tag{7.34}$$

This scheme is known as static path approximation plus random phase approximation (SPA+RPA), although the RPA is carried out, not only around the mean field σ_o, but also in configurations away from the mean field. Setting $\mathcal{C} = 1$ in Eq. (7.33), one recovers the SPA expression for the partition function.

7.3 Applications of the SPA+RPA to Level Densities

The SPA+RPA method provides a powerful tool to simultaneously deal with both large- and small-amplitude fluctuations.[1] It has been used within the

---

[1] Simple estimates of the effects quantal and thermal fluctuations have on the damping width of the GDR are discussed in Ch. 10.

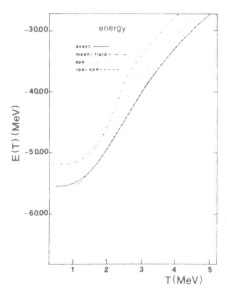

Figure 7.3. The energy as a function of the temperature for the modified Lipkin model described in the text. (From Puddu et al. (1991).)

framework of schematic models of the nucleus (cf. Puddu et al. (1991) and references therein) to calculate the energy of the system as a function of the temperature, and the level density as a function of the energy, and to compare the results obtained in the different approximations with the exact solution of the problem. Calculations for the nucleus ^{170}Er were also performed (Puddu (1993)).

7.3.1 A schematic model

Results obtained (Puddu et al. (1991)) solving the SPA+RPA equations for the case of a Lipkin-type model in comparison with the corresponding exact solutions, are displayed in Figs. 7.3 and 7.4.

The model consists of 10 particles distributed over two levels of total degeneracy 20 (half-filled shell situation), separated by an energy gap of 10 MeV, and connected by the operator $\hat{Q}$ with unit matrix elements (cf. Eq. (7.27)). The strength of the interaction $\chi = 0.6565$ is such that the system is deformed ($\sigma_0 \neq 0$) at $T = 0$ and becomes spherical ($\sigma_o = 0$) at $T > T_c = 2.5$ MeV, mimicking the shape phase transition, taking place in heavy nuclei away from closed shells, from quadrupole deformed to spherical configurations (cf. Fig. 8.6). The basic result emerging from the comparison of the exact results with those obtained in the mean-field approximation (Eq. (7.32)), in the SPA approximation (Eq. (7.31)), and in the SPA+RPA approximation (Eq. (7.33)), is that the signature of the phase transition, which is particularly clear in the

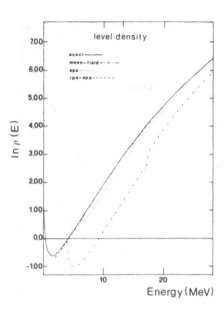

Figure 7.4. The level density as a function of the energy for the modified Lipkin model described in the text. (From Puddu et al. (1991).)

mean-field results, is smoothed by the fluctuations. Furthermore it is seen that, at high temperature the SPA reproduces the level density enhancement over the mean-field value. Such a result can be understood in terms of the discussion concerning the approximations introduced after Eqs. (7.28) and (7.29), and of the integration over values of $\bar{\sigma}$ different from $\sigma_o = 0$, which collects sizeable contributions to the level density of the states feeling shallower potentials. The change of slope of the exact and of the SPA+RPA curves shown in Fig. 7.4, is expected from the structure of the factor C in Eq. (7.34). The small-amplitude quantal fluctuation contribution tends to vanish for temperatures greater than typical single-particle excitations energies, in overall agreement with the decrease of the value of the level density parameter a found experimentally (cf. Fig. 7.2).

7.3.2 A realistic model

A model which has been extensively used to describe the single-particle and collective properties of open-shell nuclei is based on the pairing plus quadrupole Hamiltonian (cf., e.g., Bohr and Mottelson (1975) and refs. therein, cf. also Bes and Sorensen (1969)). That is, a Hamiltonian of the type given in Eq. (7.27), describing a nucleon of mass equal to the bare mass moving in a Saxon–Woods potential. The one-body operator $\hat{Q}_\nu$ is, in this model, the quadrupole operator. To this term is added a pairing force. This model was used (Puddu (1993)), to calculate the level density in the nucleus ^{170}Er within the framework of

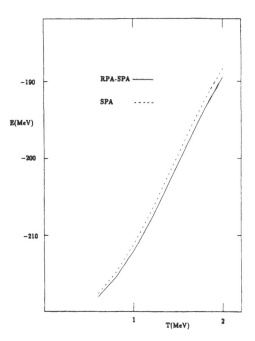

Figure 7.5. The energy as a function of the temperature for the pairing + quadrupole Hamiltonian of Eq. (7.35) in the nucleus ^{170}Er. (From Puddu (1993).)

the SPA+RPA equations discussed above. A number of 20 protons and 18 neutrons were distributed among 6 and 5 single-particle levels respectively. The results of the calculations are shown in Figs. 7.5 and 7.6. They display features which are rather similar to those obtained within the framework of the schematic model (cf. Figs. 7.3 and 7.4).

7.3.3 The Monte-Carlo shell model method

The essence of the Monte-Carlo Shell Model Method consists in calculating, making use of Monte-Carlo techniques, the functional integral for the partition function Z defined in Eq. (7.30) "exactly," that is, without recurring to any of the approximation discussed above.

To improve the accuracy, the time-interval β is split in n_t time-slices $\Delta\beta/n_t$ (cf. Ormand (1997) and Koonin et al. (1997)). Following this approach the thermal properties of the nucleus ^{54}Fe have been calculated (Dean et al. (1994)), considering 8 valence neutrons and 6 valence protons moving in the 2p–1f shell and interacting via a realistic Hamiltonian. The total number of configurations involved in the calculation of the canonical ensemble was of the order of 10^9. For a temperature $T \approx 1.1$–1.5 MeV, the value of the level

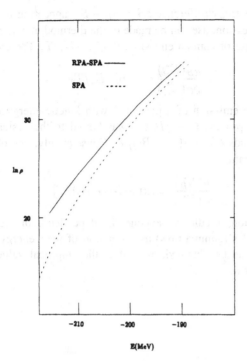

Figure 7.6. The level density as a function of the energy for the pairing + quadrupole Hamiltonian of Eq. (7.35) in the nucleus ^{170}Er. (From Puddu (1993).)

density parameter obtained was ≈ 7.5 MeV^{-1}, the experimental value being of the order of 6.8 MeV^{-1}.

7.4 Use of Level Densities

The decay rate of a compound nucleus C into a channel $B + b$, where B is the residual nucleus and b is a nucleon, a light particle or a gamma ray with wavenumber k_b and spin I_b, can be expressed as

$$\frac{d^2 N_b}{dE dt} = \frac{(2I_b + 1)k_a^2}{2\pi^2 \hbar}\sigma(b + B \to C)\frac{\rho_B(E_B^*)}{\rho_C(E_C^*)}. \tag{7.35}$$

The expression is based on detailed balance and therefore relates the decay rate for the process $C \to B + b$ to the average cross section for forming the compound state C in the inverse process. To a good approximation, the ratio of densities of states appearing in Eq. (7.35) can be expressed in terms of the entropy of the two systems as

$$\frac{\rho_B(E_B^*)}{\rho_C(E_C^*)} \approx \exp[S_B(E_B^*) - S_C(E_C^*)]. \tag{7.36}$$

To calculate the entropy difference $\delta S = S_B - S_C$ appearing in the exponent of the above expression, use can be made of the thermodynamic relation $\delta S = \delta E^*/T$. In the case of gamma emission, $\delta S = -E_\gamma/T$. Therefore

$$\frac{\rho_B(E_B^*)}{\rho_C(E_C^*)} \approx \exp[-E_\gamma/T]. \tag{7.37}$$

In the case of the emission of a particle b with kinetic energy ϵ_b and which before the decay process $C \to B + b$ was bound to the system C with an energy B_b, one finds $\delta S = -(\epsilon_b + B_b)/T$. Consequently, one obtains for the density of states ratio

$$\frac{\rho_B(E_B^*)}{\rho_C(E_C^*)} \approx \exp[-(\epsilon_b + B_b)/T]. \tag{7.38}$$

The above equations predict an exponential dependence of the spectrum of evaporated particles (gamma-rays) as a function of their energy, from which the temperature can be obtained, as well as the empirical value of the level density parameter a.

8

Linear Response

In what follows we shall consider the formal description of the response and of the correlations displayed by the nuclear many-particle system at finite temperature. In particular, we shall work out the theoretical tools which allow to describe mean field properties, of both single-particle and collective motion, as well as their interweaving in highly excited systems. We shall see that the analysis of the response of the system to a weak external probe, and the treatment of correlations proceed in much the same way as they did at zero temperature. There is, however, a main difference to be kept in mind. At finite temperature, one no longer finds a unique state which describes the system, and a statistical description is called for. To study a quantum mechanical system, it is generally convenient to use the grand canonical ensemble discussed in Sect. 7.1, for which the energy of the system and the number of particles are not required to have definite values but only particular average values.

8.1 Mean Field Theory

The derivation of the thermal Hartree–Fock equations follows closely that at $T = 0$. The resulting equation can be written as

$$H[\varrho]\varphi_\nu(\vec{r}) = \varepsilon_\nu\varphi_\nu(\vec{r}), \tag{8.1}$$

where the temperature enters through the particle density

$$Tr\varrho = \sum_j n_j = \langle \hat{N} \rangle, \tag{8.2}$$

the occupation number of the level j being defined as

$$n_j = \frac{1}{1 + \exp(\beta\varepsilon_j - \alpha)}. \tag{8.3}$$

The Hartree–Fock equations (8.1)–(8.3) at finite temperature are the result of the minimization of the thermodynamical potential Ω:

$$\delta\Omega = \delta(E - TS - \mu N) = 0. \tag{8.4}$$

The variation of the wavefunction $\varphi_j(\vec{r})$ leads to Eq. (8.1) while the variation of the occupation numbers n_j gives Eq. (8.3). In Eq. (8.4), E is the average nuclear energy $E = \langle \hat{T} + \hat{v} \rangle$, v being the two-body interaction acting among the nucleons. The entropy is defined in the standard form,

$$S = -\sum_j \{n_j \ln(n_j) + (1 - n_j)\ln(1 - n_j)\}. \tag{8.5}$$

The internal excitation energy of the system at a given temperature is defined by

$$E^*(T) = E(T) - E(0), \tag{8.6}$$

where

$$E(T) = \sum_{\nu_i \leq \nu_F} \varepsilon_i n_i - \frac{1}{2} \sum_{\nu_i, \nu_k \leq \nu_F} \langle \nu_i \nu_k | v | \nu_i \nu_k \rangle. \tag{8.7}$$

The single-particle levels obtained by solving the Hartree–Fock equations (8.1)–(8.3) are essentially constant with temperature. An example is provided by the results displayed in Fig. 8.1. In the figure, the proton energy levels of ^{208}Pb are shown as a function of temperature. Although calculations at temperatures larger than 4–5 MeV cannot be trusted in that the nucleus at these temperatures probably cannot hold together any more, and a so-called liquid–gas phase transition is likely to take place (Bertsch and Siemens (1983)), the results provide clear evidence for the stability of the mean field with respect to temperature. This constancy was expected, at least for values of T considerably smaller than that of the energy difference between major shells ($\hbar\omega_0 = 41A^{-1/3}$ MeV ≈ 7 MeV for $A = 208$). Looking in more detail at the single-particle levels, one observes that, as a rule, the energy of the last occupied level increases slightly with temperature while the lowest levels lying above the Fermi energy decrease. This result can be qualitatively understood in terms of the variation of the density with temperature. An example of this dependence is shown in Fig. 8.2. The heating of the system results in a smoothing of the shell oscillations, a decrease of the central density and a broadening of the surface region. This trend results in an increase of the root mean square radius as shown in Fig. 8.3, as well as in a more diffuse mean field potential (cf. Fig. 8.4). In other words, the system expands and thus becomes, even if slightly, less dense. Due to the fact that the non-locality of the Hartree–Fock potential is directly connected with the Pauli principle, the quantitative importance of the effect decreases with decreasing nuclear density. Because the k-mass (cf. Eq. (3.13) and Sect. 4.1.2) embodies the non-locality of the mean field, one expects that its value will progressively approach that of the bare mass as a function of the temperature (cf. Fig. 8.4). An increase of

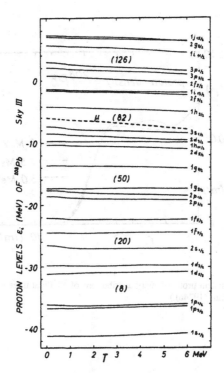

Figure 8.1. Proton levels ε_j of ^{208}Pb as a function of temperature T. The dashed curve is the chemical potential μ. (After Brack and Quentin (1974).)

the k-mass will increase the density of levels around the Fermi energy. That is, it raises the energy of levels below the Fermi energy and decreases that of the levels above the Fermi energy (cf. Sect. 3.1.1). On the other hand, an increase in the radius of the system lowers the whole single-particle spectrum by decreasing the kinetic energy of the nucleons. These two effects try to cancel each other for the occupied orbits, while they reinforce each other for the unoccupied states.

Because of the extreme constancy of the single-particle levels with temperature, one expects that the values of the different thermodynamic quantities, in particular E^* and S, will be very similar to the values

$$E^{*(0)}(T) = \sum_{\nu_i} \varepsilon_i^{(0)} n_i^{(0)} - \sum_{\nu_i \leq \nu_F} \varepsilon_i^{(0)}, \qquad (8.8)$$

and

$$S^{(0)}(T) = -\sum_{\nu}\{n_\nu^{(0)} \ln(n_\nu^{(0)}) + (1 - n_\nu^{(0)})\ln(1 - n_\nu^{(0)})\}, \qquad (8.9)$$

where $n_\nu^{(0)}$ is the Fermi occupation number associated with the single-particle levels calculated with the single-particle energies $\varepsilon_\nu^{(0)} = \varepsilon_\nu(T = 0)$. The

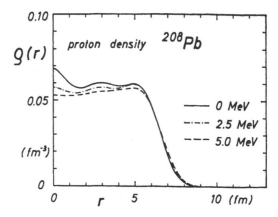

Figure 8.2. Self-consistent proton density distributions of ^{208}Pb at three different temperatures. (After Brack and Quentin (1974a).)

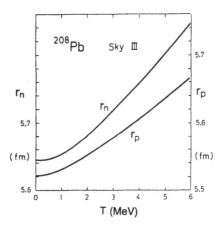

Figure 8.3. Proton and neutron root-mean-square radii of ^{208}Pb as a function of temperature. (After Brack and Quentin (1974a).)

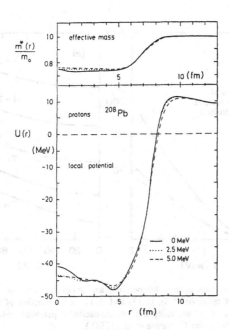

Figure 8.4.   Properties of the proton Hartree–Fock Hamiltonian of ^{208}Pb at three different temperatures. In the upper part the k-effective mass $m_k(r)$ in units of the nuclear bare mass m is shown as a function of r. In the lower part the Coulomb plus nuclear potentials are displayed. (After Brack and Quentin (1974a).)

associated proton and neutron chemical potentials are determined through the relations

$$N = \sum_{j_n} n_{j_n}^{(0)}, \quad Z = \sum_{j_p} n_{j_p}^{(0)}, \tag{8.10}$$

where j_n and j_p label neutron and proton orbitals respectively. In Fig. 8.8.5(a) the excitation energy defined in Eq. (8.8) is plotted as a function of the temperature. Making use of the relation (cf. Sect. 7.2.1)

$$E^* = aT^2, \tag{8.11}$$

it is found that a value of the order of $a = A/8$ MeV reproduces the result of the calculation.

In heavy-ion reactions leading to fusion and eventually to a compound nucleus, the system acquires, as a rule, a large amount of angular momentum. The angular momentum-dependent Fermi occupation number can be written as

$$n_j = \{1 + \exp[(\varepsilon_j - \varepsilon_f - m_j \delta)/T]\}^{-1}, \tag{8.12}$$

where m_j is the magnetic quantum number associated with the single-particle level j. The Lagrange multiplier δ is determined, for protons and neutrons, by

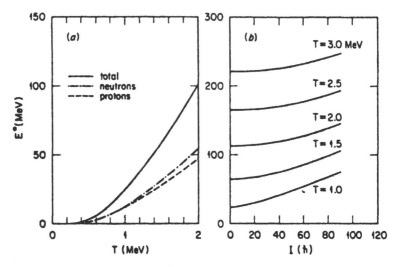

Figure 8.5. Relation between energy, temperature and angular momentum in the Pb-region, calculated making use of relations (8.8), (8.13) and (8.14), as a function of temperature. The single-particle levels were calculated using a harmonic oscillator potential with parameters taken from Nilsson et al. (1969). (After Civitarese et al. (1984).)

the conditions

$$I_p = \sum_{j_p} n_{j_p} m_{j_p},$$ (8.13)

and

$$I_n = \sum_{j_n} n_{j_n} m_{j_n},$$ (8.14)

respectively. Here, I_p and I_n are the total angular momentum associated with protons and neutrons, respectively. Assuming the total angular momentum I is distributed between protons and neutrons according to

$$I_p = \frac{Z}{A}I, \ I_n = \frac{N}{A}I,$$ (8.15)

the relation between I and E^* can be calculated. An example is reported in Fig. 8.5(b).

8.2 Shell Corrections

In nuclei with a deformed ground state, one expects larger changes of the average field with temperature than the one discussed in the previous section in connection with the closed shell system ^{208}Pb. This is because deformation depends on the polarization effects that a reduced number of nucleons outside closed shells have on the shape of the system. Essentially all of the strongly deformed nuclei display quadrupole prolate shapes in their ground state (Bohr and Mottelson (1975)). The value of the deformation parameters observed, are of the order of the ratio of the particles outside closed shells to the total number of particles. An estimate of these deformations can be obtained making use of a model in which each nucleon moves independently in the average deformed field produced by the rest of the nucleons. The equilibrium deformation can be determined from a self-consistency argument, requiring that the eccentricity of the potential and that of the density should be equal (cf., e.g., Mottelson (1962)). It then follows that important contributions to the nuclear potential energy function result from the effect of shell structure. The nonuniformities in the one-particle eigenvalue spectrum associated with the shell structure imply that the nuclear energy will not vary smoothly with particle number, as in the description of the nucleus in terms of bulk properties provided by either the liquid drop or the Fermi gas model, but will exhibit specific variations depending on the approximate degeneracies in the spectrum of single-particle motion and in the corresponding filling of the shells. The inadequacy of the liquid-drop analysis can be attributed to the long mean free path for the motion of the individual nucleons in the nucleus, opposite to the hydrodynamical hypothesis. The limitation of the Fermi gas model is connected with the fact that the mean free path is in many situations not only larger than the average distance between nucleons (cf. Eq. (4.38)) but also larger than the dimensions of the system (cf. Eq. (4.18)).

The contribution of shell structure to the potential energy function, the shell structure energy, is determined by the one-particle energies of the occupied states,

$$\mathcal{E}_{ip} = \sum_{j=1}^{A} \varepsilon_j. \tag{8.16}$$

It is to be noted that the independent-particle energy $\mathcal{E}_{ip}$ does not represent the total nuclear energy. However, the deviations of $\mathcal{E}_{ip}$ from a smooth function of particle number represents a valid measure of the shell energy (Strutinsky (1967), Nilsson et al. (1969), Brack et al. (1972)). The effects of pair correlations, which can be essential at $T = 0$, can be included by replacing ε_j by the quasi-particle energy. For a system containing many particles, the main part of $\mathcal{E}_{ip}$ varies smoothly with particle number. This part, $\tilde{\mathcal{E}}_{ip}$, can be identified by considering the asymptotic behaviour of $\mathcal{E}_{ip}$ in the limit of large particle number. The shell energy is then obtained as the difference

$$\mathcal{E}_{sh} = \mathcal{E}_{ip} - \tilde{\mathcal{E}}_{ip} \tag{8.17}$$

between $\mathcal{E}_{ip}$ and the asymptotic function $\tilde{\mathcal{E}}_{ip}$. The standard method to calculate $\tilde{\mathcal{E}}_{ip}$ is based on an energy average of the single-particle level density.

Making use of the harmonic oscillator potential to describe the single-particle motion, and assuming closed shells in both protons and neutrons, one obtains

(Bohr and Mottelson (1975))

$$\mathcal{E}_{sh} \approx -4.5 A^{1/3} \, \text{MeV}. \tag{8.18}$$

The above estimate implies a value of ≈ 27 MeV for ^{208}Pb. The empirical value is about 13 MeV. The reduction of about a factor of 2 can be related to the fact that single-particle levels within each major shell are spread over an energy interval of the order of half the separation energy between major shells. In particular, in the case of ^{208}Pb the single-particle gap associated with the shell closure is of the order of 3.5 MeV, while the separation between major shells is about 7 MeV. At finite temperatures, the effects of the shell structure in the one-particle motion are of decreasing importance, because of the many configurations that may contribute to it. For sufficiently high temperatures, the nuclear properties are expected to vary smoothly with particle number.

The analysis of a schematic, yet relevant model to describe the single-particle motion (Ericson (1958), Bohr and Mottelson (1975)) and which allows to calculate analytically the shell energy as well as the entropy, indicates that shell structure effects become quite small already for temperatures of the order of few tenths of the shell spacing. That is, $T_c \approx 3$–4 ($\hbar\omega_0/10$), with $\hbar\omega_0 \approx 41 A^{-1/3}$ MeV (i.e., $\approx \varepsilon_F A^{-1/3}$). In the case of ^{208}Pb, $T_c \approx 2.1$–2.8 MeV. Results of numerical calculations (Brack and Quentin (1974a)) are displayed in Fig. 8.6 where the free energy $F = E - TS$ for the nucleus ^{168}Yb are shown for four values of the temperature. At $T = 0$ this nucleus displays a static quadrupole distortion. For a value of $T \approx 4$ ($\hbar\omega_0/10$) ≈ 3 MeV, the deformation energy behaves like that describing the liquid drop model of the nucleus. In particular the minimum is at zero quadrupole moment.

The nuclear free energy F as a function of deformation and temperature can be written as

$$F(T, \vec{\alpha}, J) = F(T, \vec{\alpha}, \omega = 0) + \frac{(J + \frac{1}{2})^2}{2\Im(\beta, \gamma, \theta, \psi)} \tag{8.19}$$

where $F(T, \vec{\alpha}, \omega = 0)$ is the free energy evaluated in the cranked-shell model approximation with rotational frequency ω equal to zero. The variable $\vec{\alpha}$ denotes the quadrupole deformation parameters β and γ and the Euler angles ϕ, θ and ψ. According to the discussion above, one may write it as a sum of a smoothly varying function F_{ld}, the liquid-drop free energy, and a shell-correction term F_{sh}. The same is true for the moment of inertia $\Im$, written as a sum of the rigid-body value and a shell-correction value.

8.2.1 A numerical example

The numerical evaluation of the thermal average of an observable $\hat{O}$ (cf. Eq. (7.6)), requires a large number of points in the five dimensional space spanned by $\vec{\alpha}$, and a microscopic determination of F_{sh} along the lines described above, for example using the Nilsson–Strutinsky procedure (Nilsson et al. (1969),

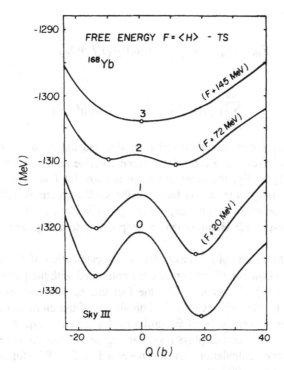

Figure 8.6. Potential (free) energy of ^{168}Yb versus mass quadrupole moment at different temperatures ($T = 0, 1, 2, 3$ MeV). The curves with $T > 0$ are shifted upwards by the amount indicated in parenthesis. Pairing correlations have been included in the calculations. (After Brack and Quentin (1974).)

Strutinsky (1967)), may be too time consuming. Therefore, it is useful to parametrize the free energy, using functions that mimic the behaviour of the Nilsson–Strutinsky calculations as closely as possible. Because the free energy is a scalar quantity, it must be a function of the two independent rotational invariants β^2 and $\beta^3 \cos(3\gamma)$ of the quadrupole deformation (Bohr and Mottelson (1975)), that is,

$$F(T, \beta, \gamma) = F_0(T) + A(T)\beta^2 - B(T)\beta^3 \cos(3\gamma) + C(T)\beta^4 + \cdots \quad (8.20)$$

Although this Landau parametrization (cf. Levit and Alhassid (1984)) gives a good overall description of the free energy, in particular regarding shape transitions, it may not be adequate for the calculation of the photoabsorption cross section in the presence of large amplitude fluctuations in the (β, γ)-plane. This is because Eq. (8.20) attempts at combining both the liquid drop free energy and the associated shell correction into the same parametrization.

In keeping with the fact that F_{sh} is expected to oscillate with deformation, being damped at large values of β, as well as to decrease with temperature, melting for $T \sim 2$–3 MeV, a possible parameterization of F_{sh} is (Ormand et al. (1996), (1997))

$$F_{sh}(\beta,\gamma,T) = \sum_{l=0}^{\text{even}} a_l j_l(b_l\beta)c_l T / \sinh(c_l T)$$

$$+ \sum_{l=3}^{\text{odd}} a_l j_l(b_l\beta)\cos(3\gamma)c_l T / \sinh(c_l T), \qquad (8.21)$$

where the j_l are the spherical Bessel function, and the a_l, b_l, c_l parameters are fitted to the results of a limited but representative number of microscopic calculations. Typically, the sums converge already for $l \approx 5,6$. The factor $c_l T / \sinh(c_l T)$ is the attenuation factor of the shell corrections with temperature. It should be noted that the expression given in Eq. (8.21) is exact in the case of a single-particle Hamiltonian, corresponding to a degenerate harmonic oscillator.

In Fig. 8.7 the results of a Nilsson–Strutinsky calculation of F_{sh} for the nucleus ^{208}Pb as a function of temperature are compared with the parametrization given in Eq. (8.21). In keeping with the fact that to calculate the quantities defined in Eq. (8.19) requires also the knowledge of the moment of inertia of the system, a parametrization of $\Im$ similar to that given in Eq. (8.21) for F_{sh} has been proposed, which displays a similar degree of accuracy in reproducing Nilsson–Strutinsky calculations as that observed in Fig. 8.7, for the case of F_{sh} (Ormand et al. (1997a)).

8.3 Sum Rules

The calculation of the sum rules at finite temperature runs parallel to that at $T = 0$, with the obvious introduction of T through the thermal probabilities given by the trace of the operator defined in Eq. (7.7). In fact, the basic double commutator introduced in Eq. (3.34) is now written as

$$\frac{1}{2}\langle 0|[\hat{F},[H,\hat{F}]]|0\rangle$$
$$= Z^{-1}\sum_n\sum_k e^{-\beta(E_n-\mu N)}\langle n|\frac{\hbar}{2m}(\vec{\nabla}_k F_k(\vec{r}))^2|n\rangle. \qquad (8.22)$$

In the case of electric dipole radiation the corresponding sum rule involves the E1 operator referred to the center-of-mass of the system (cf. Eq. (3.39) and Sect. 10.1), i.e.,

$$F(\vec{r}_k) = e\left(\frac{N-Z}{A} - t_z(k)\right)r_k Y_{1\mu}(\hat{r}_k) \qquad (8.23)$$

and the corresponding energy weighted sum rule (oscillator strength) is

$$S(E1,T) = S(E1,T=0) = \frac{9}{4\pi}\frac{\hbar^2 e^2}{2m}\frac{NZ}{A}, \qquad (8.24)$$

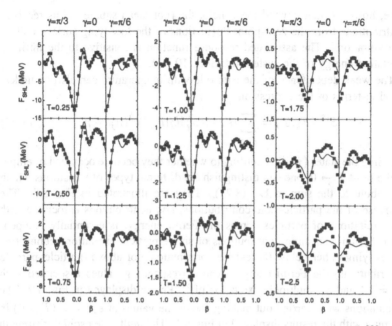

Figure 8.7. Nilsson–Strutinsky shell corrections (solid squares) to the free energy for ^{208}Pb as a function of temperature for oblate ($\gamma = \pi/3$), triaxial ($\gamma = \pi/6$), and prolate ($\gamma = 0$) shapes at zero angular momentum. The parameterization to F_{sh} given in Eq. (8.21) is represented by the solid line. (After Ormand et al. (1997).)

an expression which coincides with the result at $T = 0$ (cf. Eq. (3.40)). The independence with T of the energy weighted dipole sum rule is in keeping with the fact that this quantity depends only on the number and on the mass and charge of the nucleons.

For $\lambda \geq 2$, the electric sum rules read

$$S(E\lambda, T) = \frac{\lambda(2\lambda + 1)^2}{4\pi} \frac{\hbar}{2m} Ze^2 \langle r^{2\lambda-2} \rangle_T,\qquad(8.25)$$

where

$$\begin{aligned}\langle r^{2\lambda-2} \rangle_T &= \sum_n \frac{e^{-\beta(E_n - \mu)}}{Z} \langle n | r^{2\lambda-2} | n \rangle \\&\approx \frac{\int \varrho_0(T) r^{2\lambda-2}\, d^3r}{\int \varrho_0(T)\, d^3r} = \frac{\int \varrho_0(T) r^{2\lambda-2}\, d^3r}{A}.\end{aligned}\qquad(8.26)$$

One expects a dependence of the sum rule with temperature, related to the fact that the nucleus expands with T. This dependence, however, cannot be very large, due to the fact that the nucleus is very rigid. This can be seen from Fig. 8.3, where the mean square radius calculated in the Hartree–Fock approximation is shown as a function of temperature. In the interval $0 \leq T \leq 3$ MeV, the mean square radius of both protons and neutrons changes by 1–2%.

Note, however, that beyond mean field, the root mean square radius receives contributions from the zero-point fluctuations of the low-lying surface collective vibrations. The associated response function is sensitive to the intrinsic excitation energy of the system (cf. Sect. 10.4.1).

The weak dependence of the sum rule with temperature can also be understood in terms of the expression

$$S^{(0)}(F_\lambda, T) = \sum_{j \geq j'} (\varepsilon_j - \varepsilon_{j'}) |\langle j|F|j'\rangle|^2 (n_{j'} - n_j). \qquad (8.27)$$

Labeling the states (j, j') according to whether they become occupied or empty in the limit $T \to 0$, one can distinguish six different types of transitions which contribute to the sum in Eq. (8.27). They are illustrated in Fig. 8.8. The decrease of the particle–hole contributions, i.e., contributions associated with the excitations of particles across the Fermi energy, is essentially compensated by non-vanishing contributions of single-particle excitations connecting levels lying either below (hole–hole contributions) or above (particle–particle contributions) the Fermi energy. The corresponding values associated with $\lambda^\pi = 2^+$ and 3^- multipole excitation of ^{208}Pb are displayed in Fig. 8.9. The calculations were carried out making use of the values of $\varepsilon_j^{(0)} = \varepsilon_j(T = 0)$ in keeping with the results displayed in Fig. 8.1. The multipole oscillator strength hardly changes in the interval $0 \leq T \leq 3$ MeV.

In Fig. 8.10 the unperturbed strength functions

$$S^{(0)}(E) = \sum_{\nu_k, \nu_i} |\langle \nu_k \nu_i^{-1}|F_{\lambda\mu}|0\rangle|^2 \delta(E - (\varepsilon_{\nu_k} - \varepsilon_{\nu_i})), \qquad (8.28)$$

for quadrupole and octupole electric fields are displayed as a function of the energy E, and for two values of the temperature. The calculations were carried out for the nuclei ^{208}Pb and ^{114}Sn and are displayed in histogram form in bins of 1 MeV. The response function for the free fermion system is not only independent of T, in terms of the total sum rule, but even the detailed distribution of strength remains almost unaffected as the temperature increases.

8.4 Random Phase Approximation

In Sect. 3.2.1 the Random Phase Approximation was derived in terms of the particle-vibration coupling. The basic assumption of the model is that the vibrations have small amplitude ($\alpha^2 \ll \alpha$), that is, that they are harmonic. The same results can be obtained by diagonalizing the particle-vibration coupling Hamiltonian $H_{PV} = T + U(r, R)$, where $U(r, R)$ has been defined in Eq. (3.15), treating it as an harmonic oscillator Hamiltonian, that is,

$$[H_{PV}, \Gamma_\alpha^\dagger] = \hbar\omega_\alpha \Gamma_\alpha^\dagger. \qquad (8.29)$$

Here $\Gamma_\alpha^\dagger$ is the creation operator of a phonon. Because of the boson character of the operator $\Gamma_\alpha^\dagger$ (cf. Eq. (3.32)), the only Hamiltonian which fulfills the

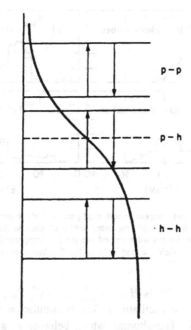

Figure 8.8. All possible positive and negative frequency transitions which can be induced by a one-particle field in a Fermi system at finite temperature. The (particle–particle)-like transitions are indicated by the label p–p, while the (particle–hole) and (hole–hole)-like transitions are labeled p–h and h–h, respectively. (After Civitarese et al. (1984).)

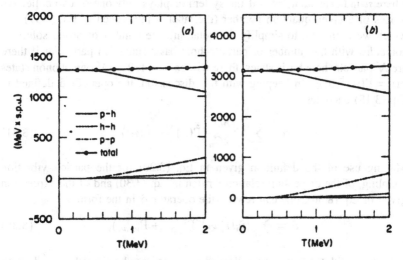

Figure 8.9. The quadrupole and octupole energy weighted sum rules for ^{208}Pb, as a function of temperature. For all temperatures displayed in the figure, the particle–hole contribution (cf. Fig. 8.8) plays a more important role than the hole–hole and particle–particle contributions because of phase space reasons. (After Civitarese et al. (1984).)

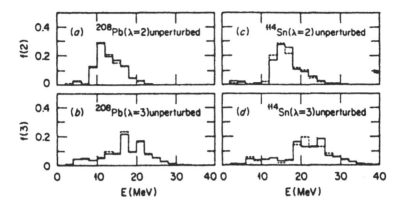

Figure 8.10. Quadrupole and octupole unperturbed particle–hole response function of ^{208}Pb and ^{114}Sn. The fraction of the total EWSR associated with each particle–hole excitation is displayed in bins of 1 MeV. The calculations were carried out for two temperatures, namely, $T = 0$ MeV (continuous line) and $T = 2$ MeV (dashed line). (After Civitarese et al. (1984).)

relation given by Eq. (8.29) is $H_{osc} = \sum_{\alpha'} \hbar\omega_{\alpha'}\Gamma_{\alpha'}^{\dagger}\Gamma_{\alpha'}$, that is, the Hamiltonian describing harmonic oscillations. The commutation relation in Eq. (8.29) selects that part of the Hamiltonian which behaves as a harmonic oscillator and diagonalizes it. The boson creation operator is defined as

$$\hat{\Gamma}(n)^{+} = \sum_{\nu_k,\nu_i} X^n(\nu_k,\nu_i)\hat{\Gamma}_{\nu_k,\nu_i}^{\dagger} + Y^n(\nu_k,\nu_i)\hat{\Gamma}_{\nu_k,\nu_i}, \qquad (8.30)$$

where n indicates that, even if the system displays only one or two collective modes of a given quantum number (the label α introduced in Eq. (3.31) and which we omit here to simplify the notation), the number of boson solutions coincides with the number of particle–hole basis states. In particular if there are $\mathcal{N}$ particle–hole basis states, there will be $n = 1, 2 \ldots \mathcal{N}$ one phonon states $|n\rangle = \hat{\Gamma}(n)^{+}|0\rangle_B$. In keeping with this discussion, the operator $\hat{\alpha}$ defined in Eq. (3.18) becomes

$$\hat{\alpha} = \sum_n \sqrt{\frac{\hbar\omega_n}{2C_n}}(\hat{\Gamma}(n)^{+} + \hat{\Gamma}(n)). \qquad (8.31)$$

Making use of the definition given in Eq. (3.20) for the particle-vibration coupling, of the dispersion relation written in Eq. (3.30) and of the expression given in Eq. (8.30), one can rewrite the operator $\hat{\alpha}$ in the form

$$\hat{\alpha} = \sum_{\nu_k,\nu_i} \langle \tilde{\nu}_i|\hat{F}|\nu_k\rangle(\hat{\Gamma}_{\nu_k,\nu_i}^{\dagger} + \hat{\Gamma}_{\nu_k,\nu_i}), \qquad (8.32)$$

in keeping with the fact that a collective excitation can be viewed, as a vibration of the nuclear surface, or as a linear combination of particle–hole excitations. Now, the above expression is nothing else but the second quantization expression of the one-particle operator $\hat{F}$. Consequently, the particle-vibration coupling term δU can be also viewed as a two-body interaction $\delta U = -\kappa \hat{F}^\dagger \hat{F}$. The particle–vibration coupling Hamiltonian can then be written as (cf. Eq. (3.41))

$$\hat{H}_{PV} = \hat{H}_0 - \kappa \hat{F}^\dagger \hat{F}, \tag{8.33}$$

with

$$\hat{F} = \sum_{\nu_k, \nu_i} \langle \tilde{\nu}_i | F | \nu_k \rangle (\hat{\Gamma}^\dagger_{\nu_k \nu_i} + \hat{\Gamma}_{\nu_k \nu_i}). \tag{8.34}$$

In Eq. (8.33) the operator $\hat{H}_0 = \sum_\nu \varepsilon_\nu a_\nu^\dagger a_\nu$ is the single-particle Hamiltonian, ε_ν being the Hartree–Fock single-particle energies.

8.4.1 Dispersion relation

Making use of the fact that

$$\langle [\Gamma_{\nu_k, \nu_i}, \Gamma^\dagger_{\nu_k, \nu_i}] \rangle = (n_i - n_k),$$

where n stands for the Fermi occupation numbers, and that the thermal average of the fermion anticommutator is

$$\{a_\nu, a_k^\dagger\} = \delta(\nu, k),$$

Eq. (8.22) leads to

$$\sum_{\nu_k, \nu_i} \frac{2(\varepsilon_k - \varepsilon_i) |\langle \tilde{\nu}_i | \hat{F} | \nu_k \rangle|^2 (n_i - n_k)}{(\varepsilon_{\nu_k} - \varepsilon_{\nu_i})^2 - (\hbar \omega_n)^2} = \frac{1}{\kappa}. \tag{8.35}$$

This dispersion relation essentially coincides with Eq. (3.30), but for the Fermi occupation numbers, which carry the dependence on temperature. In keeping with this result, the particle-vibration coupling strength is, in the present case, given by the relation

$$\Lambda_n^2 = \{2\hbar \omega_n \sum_{\nu_k, \nu_i} \frac{2(\varepsilon_{\nu_k} - \varepsilon_{\nu_i}) \langle \tilde{\nu}_i | \hat{F} | \nu_k \rangle^2 (n_i - n_k)}{[(\varepsilon_{\nu_k} - \varepsilon_{\nu_i})^2 - (\hbar \omega_n)^2]} \}^{-1}, \tag{8.36}$$

while the X and Y amplitudes have the same expression as those given by Eq. (3.29).

8.4.2 The collective response

The weak dependence of the energy weighted sum rule with temperature, observed in the study of the unperturbed response (cf. Fig. 8.9) is also found in the case of RPA, as can be seen in Fig. 8.11. On the other hand, the situation concerning the similarities and differences between unperturbed and correlated response functions is more subtle. In fact, in this case one has to distinguish between low-lying and high-lying modes, between isoscalar and

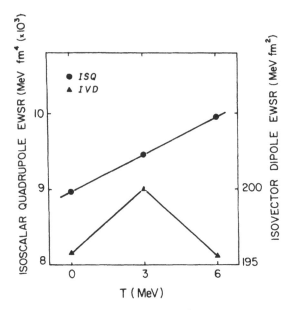

Figure 8.11. Dipole and quadrupole energy weighted sum rule associated with ^{40}Ca calculated in RPA as a function of temperature. (After Sagawa and Bertsch (1984).)

isovector responses, and whether the states belong to closed shell (normal) nuclei or open shell (superfluid) nuclei where pairing plays an important role. The variety of situations are exemplified in Figs. 8.12–8.14.

Low-lying isoscalar modes can be viewed as surface vibrations, and are expressions of the plastic behaviour of the system. It is then natural that their properties, in particular their excitation energy, depends on whether the system is closed shell or not. While in the closed-shell nuclei ^{208}Pb and ^{16}O, quadrupole surface vibrations appear at 4.1 MeV and 6.9 MeV, respectively, in the superfluid spherical nuclei ^{98}Mo and ^{114}Sn, low-lying 2^+ states are found already at about 1 MeV. This is in keeping with the fact that closed shell systems are much stiffer than open shell nuclei. Consequently, temperature brings down intensity to lower energies in closed shell nuclei while it has essentially the opposite behaviour in open shell nuclei (superfluid nuclei, cf. Figs. 8.12–8.14). In particular, in the case of ^{114}Sn the collectivity of the low-lying 2^+ surface vibration is strongly reduced by thermal effects while in the case of ^{98}Mo it is essentially obliterated. The situation is similar in the case of the octupole vibrations, although less dramatic (cf. Fig. 8.15). In other words, the collectivity of low-lying vibrations, is strongly reduced by temperature.

On the other hand, if one were to look at the low-energy spectrum with little energy resolution, a number of quantities remain rather unaltered. In fact both the fraction of the energy weighted sum rule and of the non-energy weighted

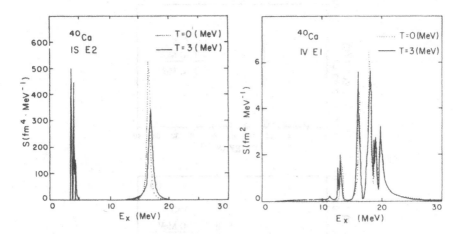

Figure 8.12. The RPA strength distribution of isoscalar (IS) quadrupole operator and of isovector (IV) dipole operator in ^{40}Ca. The dotted curve shows the result of $T = 0$ MeV, while the solid curves show the result at $T = 3$ MeV. (After Sagawa and Bertsch (1986).)

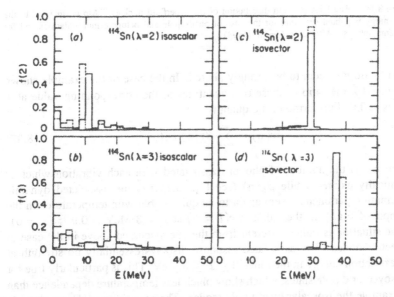

Figure 8.13. Fraction of the isoscalar and isovector energy weighted sum rule calculated in RPA as a function of the excitation energy for the closed shell nucleus ^{208}Pb and the superfluid nucleus ^{114}Sn. Both quadrupole and octupole response functions in bins of 1 MeV are shown. The solid lines correspond to the results for $T = 0$ MeV while the dashed lines express the results for $T = 2$ MeV. (After Civitarese et al. (1984).)

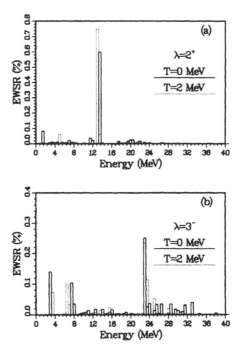

Figure 8.14. The RPA strength distribution of the superfluid nucleus ^{98}Mo normalized to the total energy weighted sum rule for the case of quadrupole and octupole multipolarities and for two temperatures. (After Donati et al. (1994).)

sum rule do not seem to be strongly altered. In the case of the example shown in Fig. 8.15, the ratio of these two quantities to the corresponding value at T = 0, is ≈ 0.6. Furthermore, the quantity

$$c(L) = \sum_n \beta_L^2(n) n_B(n) \tag{8.37}$$

where n_B is the number of phonons associated with each vibration which is thermally excited while $\beta_L(n)$ (see Eq. (3.23)) is the associated dynamic deformation parameter, seem again to be quite stable with temperature. In the example of Fig. 8.15 the value of $c(L = 3)$ at $T = 3$ MeV is $0.9\beta_3^2$ ($T = 0$).

The situation is rather different from the one discussed above in the case of isovector modes. The associated response displays essentially no strength at low excitation energy in the range $0 \leq T \leq 3$ MeV. This is particularly true for the isovector dipole states, which show much less temperature dependence than for example the isoscalar quadrupole modes. The sensitivity of the quadrupole strength function may be understood in part in terms of the softness of the nucleus, which in turns depends on the momentum distribution function of the particles (Sagawa and Bertsch (1984)). With a relatively sharp Fermi surface as in a closed shell nucleus, a quadrupole deformation implies a corresponding

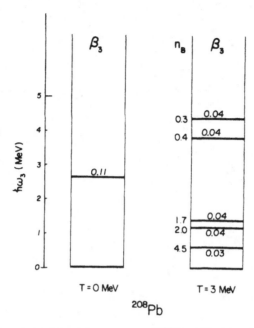

Figure 8.15. The calculated low-lying spectrum of ^{208}Pb at $T = 0$ and $T = 3$ MeV. The dynamic deformation parameters β_3 as well as the boson occupation factors are also shown.

deformation of the momentum distribution, generating a large restoring force. In an open shell nucleus, corresponding, e.g., to a closed shell system at finite temperature, deformations can occur by transitions within a major shell and the strength function has peaks at low excitation energy. On the other hand, the dipole mode is the displacement of neutrons against protons in coordinate space without any change in momentum space. Consequently, the melting of the Fermi surface is not important as far as the excitation energy and the strength distribution of the isovector dipole vibrations is concerned.

9

Collisions

Particles moving in levels closed to the Fermi energy have, in cold nuclei, a mean free path which is larger than the nuclear dimensions. In fact, due to the discrete nature of the low-energy nuclear spectrum, no real transitions can take place connecting those single-particle states to more complicated configurations. The situation is different when the system becomes internally excited, in keeping with the fact that temperature can be viewed as the energy available to the nucleons to make real transitions into complex configurations, in particular into "doorway" configurations, where the particle has changed its state of motion by setting the nuclear surface into vibration. Because of the central role played by these configurations in the relaxation of single-particle motion, it turns out that the associated damping width displays a linear dependence with temperature. Calculations show that even this weak dependence cancels out from the damping width of giant resonances, due to quantal interference between particle- and hole-contributions.

9.1 The Dynamical Shell Model

When calculating the single-particle Green's function, that is, the time evolution of a fermion in a nucleus at finite temperature, one must average over all possible configurations of the system. This is achieved through the thermal average

$$\frac{Tr[e^{-\beta H} a_\nu(t) a_\nu^\dagger(t')]}{Tr(e^{-\beta H})}, \tag{9.1}$$

where

$$a_\nu(t) = e^{+iHt} a_\nu e^{-iHt}, \tag{9.2}$$

while

$$TrA = \sum_n \langle n|A|n \rangle,$$

corresponds to a complete sum over the eigenstates of the Hamiltonian. We shall not diagonalize the total Hamiltonian

$$H_{PV} = H_0 + \delta U,$$

but shall treat

$$H_0 = T + U(r, R_0),$$

exactly, while solving perturbatively the term δU (cf. Eq. (3.16)).

To calculate the expression given in Eq. (9.1), one has to expand both $\exp(-\beta H)$ and $\exp(+iHt)$. It is of course more convenient to expand on a single quantity as done in Sect. 4.1. By considering β and t as the real and imaginary part of a complex temperature, one can regain the simplicity of the time independent calculations, having to deal with the expansion of only one exponential. This is known as Matsubara's method (cf., e.g., Mahan (1981)). It turns out that even calculations at zero temperature are more easily performed using this method than standard perturbation theory. This is because all possible processes at a given order of perturbation are included by calculating a single diagram (cf., e.g., Fig. 9.1).

9.1.1 Single-particle width: Matsubara formalism

We shall calculate the self-energy of a particle, arising from its coupling to a surface vibration. Care has to be taken of the occupation factors of the fermions as well as of the fact that the presence of thermally excited bosons is controlled by the corresponding boson occupation numbers. Because the fermion occupation factor $(1 + \exp(\beta\xi_j))^{-1}$ (with $\xi_j = \varepsilon_j - \mu$) has poles at $\xi_j = (2n + 1)i\pi/\beta$ and the boson occupation factor $(\exp(\beta\omega_\lambda) - 1)^{-1}$ has poles at $\omega_\lambda = 2ni\pi/\beta$, they can be written, aside from additive constants and overall multiplicative factors as

$$\sum_n \frac{1}{ip_n - \xi_p} \quad \text{and} \quad \sum_n \frac{1}{i\omega_n - \omega_q}, \tag{9.3}$$

where the frequencies at the pole are

$$p_n = \frac{(2n + 1)\pi}{\beta} \quad \text{fermions}, \tag{9.4}$$

and

$$\omega_n = \frac{2n\pi}{\beta} \quad \text{bosons}. \tag{9.5}$$

The index n takes all possible integer values from $-\infty$ to $+\infty$. The factors inside the summations in Eq. (9.3) have the nature of the energy denominator associated with the propagation of fermions and bosons (cf., e.g., Eq. (4.1)), whether the system is at zero or at finite temperature. Temperature information is present in the relations given in Eq. (9.3) through the frequencies (9.4)–(9.5).

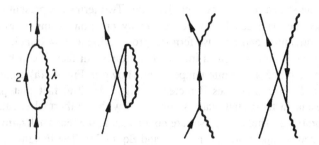

Figure 9.1. The four time-ordered $T \neq 0$ processes corresponding to the diagrams of Fig. 4.2(a) at $T = 0$.

In keeping with this discussion, the self-energy of the fermion calculated in second order perturbation theory in the particle-vibration coupling Hamiltonian (cf. Figs. 4.2(a) and 9.1) is

$$(\Sigma^{(1)}(1, ip_n))_p = -\frac{1}{\beta} \sum_{2,\lambda} V^2(1, 2; \lambda)$$

$$\times \sum_{i\omega_n} \frac{1}{i(p_n - \omega_n) - \varepsilon_2} \left(\frac{1}{i\omega_n - \omega_\lambda} - \frac{1}{i\omega_n + \omega_\lambda} \right), \qquad (9.6)$$

where i labels a particle state with energy ε_i, λ labels an RPA vibration with energy ω_λ, and $V(i, j; \lambda)$ is the particle-vibration matrix element (cf. Eq. (3.22)). The explicit linear dependence with temperature is cancelled by the boson sum $\sum_{i\omega_n} \sim \beta$. This summation is performed as described in the textbooks (Mahan (1981)) to give the following formula for the self-energy

$$(\Sigma_{ret}^{(1)}(\varepsilon + i\eta))_p = \sum_{2,\lambda} V^2(1, 2; \lambda) \left\{ \frac{1 + n_B(\lambda) - n_F(2)}{\varepsilon + i\eta - \varepsilon_2 - \omega_\lambda} + \right.$$

$$\left. \frac{n_B(\lambda) + n_F(2)}{\varepsilon + i\eta - \varepsilon_2 + \omega_\lambda} \right\}. \qquad (9.7)$$

Here $n_B(\lambda) = (\exp(\beta\omega_n) - 1)^{-1}$ and $n_F(i) = (\exp(\beta(\varepsilon_i - \mu)) + 1)^{-1}$ are the Bose and Fermi occupation numbers. In deriving the above expressions, relations of the type $n_B(ip_n - \varepsilon_2) = n_F(\varepsilon_2) - 1$ connecting these two factors have been used. Furthermore, the quantity ip_n has been replaced by $\varepsilon + i\eta$. This analytic continuation yields the retarded function, which is what one needs to compare the calculations with the experimental findings.

We now examine the imaginary part of the self-energy defined in Eq. (9.7) for the value of the averaging parameter which goes to zero, finding (Bortignon et al. (1986))

$$\lim_{\eta \to 0} Im \left(\Sigma_{ret}^{(1)}(\varepsilon + iI) \right)_p = 2\pi \sum_{2,\lambda} V^2(1, 2; \lambda)$$

$$\times \{ [1 + n_B(\lambda) - n_F(2)] \delta(\varepsilon - \varepsilon_2 - \omega_\lambda)$$

$$+ [n_B(\lambda) + n_F(2)] \delta(\varepsilon - \varepsilon_2 + \omega_\lambda) \}, \qquad (9.8)$$

which has the structure of Fermi's golden rule. Two terms are present because the intermediate state can be formed either by creation or annihilation of a vibrational quantum, and because fermion propagation at $T \neq 0$ includes both particle and hole propagation in intermediate states. In fact, it can be shown that the single process, at finite temperature, shown in Fig. 4.2(a) corresponds to the four $T = 0$ processes depicted in Fig. 9.1. The fact that phonons can be present in the initial state is connected with the thermal excitation of surface vibrations. These processes are not present in the renormalization of the particle at zero temperature (cf. Fig. 4.2 and Eq. (4.3)). The imaginary part of these graphs at finite temperature should include, besides the δ-function which ensures energy conservation, a thermal Pauli blocking factor for the particle (or hole) in the intermediate state, and a matrix element for the creation or annihilation of the vibration, with the dependence on the number of quanta of the initial state. The boson and fermion occupation numbers associated with the four terms are given by

$$(1 - n_F(2))(1 + n_B(\lambda))\delta(\omega - (\varepsilon_2 + \omega_\lambda))$$
$$+n_F(2)(1 + n_B(\lambda))\delta(\omega - (2\omega - \varepsilon_2 + \omega))$$
$$+(1 - n_F(2))n_B(\lambda)\delta(\omega + \omega_\lambda - \varepsilon_2)$$
$$+n_F(2)n_B(\lambda)\delta(\omega + \omega_\lambda - (2\omega - \varepsilon_2)). \tag{9.9}$$

The sum of these terms coincides with the expression of the occupation factors appearing in Eq. (9.8). The appearance of boson occupation factors in the above expressions is a consequence of treating all vibrations as phonons, as done in the Random Phase Approximation.

From the knowledge of the real and imaginary parts of the self-energy given in Eq. (9.6) it is possible to construct the single-particle strength function defined in Eq. (4.9). Examples associated with single-particle levels around the Fermi energy and for different values of T are displayed in Fig. 9.2. The corresponding full width at half maximum, that is, the damping width of a single-particle state at the Fermi energy as a function of temperature is shown in Fig. 9.3. The results can be expressed in terms of the sum of the value of $\Gamma_{sp}^\downarrow$ at $T = 0$ MeV, plus a linear function of T as

$$\Gamma_{sp}^\downarrow = \alpha + \beta T. \tag{9.10}$$

The value of α is strongly dependent on whether one is dealing with open or closed shell nuclei. In the first case, particles at the Fermi energy are associated with quasiparticle energies of the order of the pairing gap (≈ 1–1.5 MeV). This is the reason why $\Gamma_{sp}^\downarrow$ can be finite even at $T = 0$. This is not the case for closed shell nuclei, where a nucleon close to the Fermi energy has a mean free path considerably larger than the nuclear dimension.

Combining the results of calculations carried out both for superfluid and closed shell nuclei at both T = 0 and T$\neq$ 0 (Donati et al. (1996), Bertsch et al. (1979), Bortignon et al. (1986), Esbensen and Bertsch (1984) and (1984a);

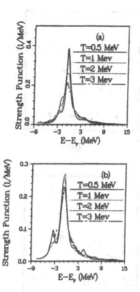

Figure 9.2. Average strength function associated with single-particle levels close to the Fermi level for a number of temperatures. In (a) the results corresponding to the nucleus ^{98}Mo are shown while in (b) are displayed those associated with ^{64}Zn. (After Donati et al. (1996).)

cf. also Sect. 4.1), a simple parametrization emerges, namely,

$$\Gamma_{sp}^{\downarrow} = 0.4\omega \quad (T = 0), \tag{9.11}$$

where $\omega = |\varepsilon - \varepsilon_F|$, in overall agreement with the experimental findings (cf. Eq. (4.17) and Fig. 4.5). On the other hand, for a particle at the Fermi energy, one obtains (cf. also Giovanardi (1996) and Giovanardi et al. (1996))

$$\Gamma_{sp}^{\downarrow} = 0.9T \quad (\varepsilon = \varepsilon_F). \tag{9.12}$$

That is, at T = 0 $\Gamma_{sp}^{\downarrow}$ displays a linear dependence with the single-particle energy measured with respect to the Fermi energy, while for a particle at the Fermi energy of a hot nucleus, $\Gamma_{sp}^{\downarrow}$ increases linearly with T. The above relations are not supposed to be used for both ω, and T different from 0 (cf. Fig. 9.3 below). Expressions (9.11) and (9.12) are quite different from the ones expected in the case of infinite systems, where $\Gamma_{sp}^{\downarrow}$ is found to depend quadratically on temperature (Landau (1957), Morel and Nozières (1962)). The reason for this difference is associated with the prominent role the surface plays in the damping of single-particle motion in nuclei (cf. also Sect. 9.3.3).

Making use of the relation given in Eq. (9.11) (cf. also Sect. 4.1), one can then calculate Γ_μ, that is, the compound nucleus damping width (cf. also Sects. 10.7 and 11.3). For this purpose use is made of the fact that the average number of quasiparticles excited in a compound nucleus at the temperature T is $v = g_0 T$. Consequently, the damping width associated with a typical compound nucleus state consists of v uncorrelated quasiparticles, each lying

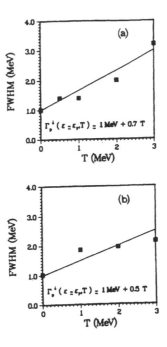

Figure 9.3. Full width at half maximum of the strength functions shown in Fig. 9.2 as a function of temperature. The results corresponding to ^{98}Mo and ^{64}Zn are shown in (a) and (b), respectively. (From Donati et al. (1996).)

at a heat energy U/v (cf. Eq. (5.12)) from the Fermi energy is

$$\Gamma_\mu(U) = v\Gamma^\downarrow_{sp}(U/v) \approx 0.5U. \tag{9.13}$$

The fact that at temperature T the Fermi energy is not sharp will induce a small correction on the result written above, correction which we shall neglect.

9.1.2 Effective mass (ω-mass)

The ω-mass was already introduced in Chapter 3 in connection with the discussion of the properties of single-particle states at zero temperature. The calculation of the ω-mass at finite temperature can be carried out again making use of the expression given in Eq. (4.24) and the real part of Eq. (9.7). The results are found (Donati et al. (1994)) to display a marked dependence with temperature which can be parametrized according to

$$m_\omega(T) = m\left(1 - \left(\frac{d\,Re\,\Sigma}{d\omega}\right)_{\epsilon_F}\right) = m + \Delta m_\omega(T), \tag{9.14}$$

where

$$\Delta m_\omega(T) = \Delta m_\omega(0)\exp(-T/T_0), \tag{9.15}$$

with

$$\Delta m_\omega(0) = m_\omega(0) - m,$$

and $T_0 \approx 2$ MeV. This result implies that already at a temperature of $T = 2$ MeV, the increase of the ω-mass over the bare value, as measured by the quantity $-\frac{\partial \Delta E(\omega)}{\partial \omega} = \Delta m_\omega(T)/m$, is reduced by $1/e$. Such a reduction is expected to have measurable consequences on a number of nuclear properties. Foremost, on the value of the density of levels parameter a, a quantity which is expected to decrease with temperature as the effective mass does ($a \sim m^*$; cf. Eq. (7.14) and subsequent discussion), in overall agreement with the experimental findings[1] (cf. Fig. 7.2). The result given in Eq. (9.15) is also expected to affect a number of properties of giant resonances, in particular the centroid of the giant dipole resonance (cf. Sect. 10.5).

9.2 Relaxation of Giant Vibrations

The Matsubara techniques can also be applied to the calculation of the self-energy of vibrations arising from the coupling of these modes to "doorway states" containing uncorrelated particle-hole excitations and a surface vibrational mode (cf. Fig. 4.12). The expression corresponding to the vertex correction (cf. Figs. 4.12(c) and (d)) reads (Bortignon et al. (1986))

$$(\Sigma^{ret}(GR, i\omega_n)_{vertex} = \frac{1}{\beta^2} \sum_{1,2;3,0;\lambda} V(1,0;GR)V(2,3;GR)$$

$$V(1,2;3,0;\lambda) \times \sum_{i\omega_n, ip_n} \{\frac{1}{i(p_n + \omega_n) - \varepsilon_1}$$

$$\times \frac{1}{i(p_n + \omega_n - \omega_{n'}) - \varepsilon_2} \frac{1}{i(p_n - \omega_{n'}) - \varepsilon_3}$$

$$\times \frac{1}{ip_n - \varepsilon_0}(\frac{1}{i\omega_{n'} - \omega_\lambda} - \frac{1}{i\omega_{n'} + \omega_\lambda})\}, \tag{9.16}$$

where the sum over the poles of the different fermion and boson occupation numbers is apparent. Similar expressions are obtained for the process where

---

[1] While the behaviour of $m_\omega(T)$ for $T_0 = 2$ MeV, when inserted in Eq. (7.14) predicts an increase of A/a for $T \approx 2$ MeV, the experimental findings seem to indicate that the transition from the $\omega-$ to the $k-$ mass regime takes place at $T_0 \approx 4$ MeV (cf. Fig. 7.2). Hot, compound nuclei display large shape fluctuations and rotate at large frequencies. The corresponding thermal shape ensemble is thus formed predominantly by deformed nuclei, which are stiffer than the spherical nuclei used in the calculations and leading to the results discussed above ($T \approx 2$ MeV). This is because the surface collectivity associated with the quadrupole degree of freedom is now tied up to rotational motion. Stiffer nuclei will lead to larger values of T_0.

the particle or the hole in the intermediate state are renormalized (cf. Figs. 4.12(a) and (b)).

The main effects of temperature on the response of giant resonances is expected to arise through the boson occupation factors. It is correct that the fermion occupation factors also vary with temperature, but the specifics of occupation probabilities is unimportant for the giant vibrations, as they affect levels around the Fermi energy within a range of the order of $T \ll \hbar\omega_0$, where $\hbar\omega_0$ are typical energies connected with particle–hole excitations associated with the elastic response of the nucleus. Changes in the fermion occupation number affect, on the other hand, in an important way the properties of the low-lying surface vibrations (cf. Figs. 8.12–8.14) which act as doorway states in the renormalization process. This effect is, of course, taken care of in terms of the particle-vibration coupling matrix element V and of the RPA energies ω_λ. Fermi occupation numbers can become important for giant resonances by increasing Landau damping (as it is in the case of low-lying vibrations), the energy scale in this case being $\hbar\omega = 41A^{-\frac{1}{3}}$ MeV, that is, the separation between major shells. Consequently, a reasonable approximation to the finite temperature damping might be obtained by keeping only the temperature dependence of the boson occupation factors, and keeping only the poles associated with the vanishing of intermediate energy denominators, that is, energy denominators resulting from the difference between the energy of the giant resonance and that of the "doorway states,"

$$\Delta\omega_{\pm\lambda,j,j'} = \omega + iI - (\pm\omega_\lambda + \varepsilon_j - \varepsilon_{j'}).$$

The corresponding expressions for the processes depicted in Figs. 4.12(d) and 4.12(a) are

$$(\Sigma^{ret}(GR, \omega + iI))_{vertex} =$$

$$-\sum_{1,2,3,4,\lambda} \frac{V(1,0;GR)V(1,2;3,0;\lambda)V(2,3;GR)}{\Delta\omega_{01}\Delta\omega_{32}}$$

$$\times \left\{ \left[\frac{n_B(\lambda) + 1}{\Delta\omega_{\lambda 20}} + \frac{n_B(\lambda) + 1}{\Delta\omega_{\lambda 13}} \right] + \left[\frac{n_B(\lambda)}{\Delta\omega_{-\lambda 20}} + \frac{n_B(\lambda)}{\Delta\omega_{-\lambda 13}} \right] \right\} \qquad (9.17)$$

and

$$(\Sigma_{ret}(GR, \omega + iI))_{self-en} = \sum_{1,0} \frac{V^2(1,0;GR)}{(\Delta\omega_{01})^2} \sum_{1,2,\lambda} V^2(1,2;\lambda)$$

$$\left(\frac{n_B(\lambda) + 1}{\Delta\omega_{\lambda 20}} + \frac{n_B(\lambda)}{\Delta\omega_{-\lambda 20}} \right). \qquad (9.18)$$

The expressions above contain the processes in which a phonon is created in the intermediate state with a factor $(n_B(\lambda) + 1)$ as expected. In addition, a physical state may be created by annihilating a phonon already present. This appears in terms proportional to $n_B(\lambda)$.

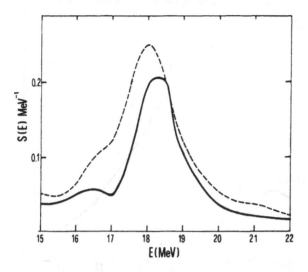

Figure 9.4. Strength function in the interval region of the main peak of the GDR of ^{90}Zr at $T = 0$ (dashed line) and at $T = 3$ (full line). (From Bortignon et al. (1986).)

In Figs. 9.4 and 9.5 the strength function

$$S_a(E) = \frac{S_0}{\pi} \frac{\frac{1}{2}\Gamma_a(E + iI) + I}{[E_a - E - \Delta E_a(E + iI)]^2 + [\Gamma_a(E + iI)/2 + I]^2}, \quad (9.19)$$

associated with the giant dipole resonance of ^{90}Zr and the giant quadrupole resonance of ^{208}Pb at two values of the temperature, are displayed (in the above expression, ΔE_a and Γ_a are the real and imaginary part of the self-energies of giant resonances). As seen from the figures, the position of the giant resonances centroid have hardly changed with temperature, and the widths have become, if anything, smaller. This result is somewhat unexpected in terms of the simple model for the damping of giant resonances discussed in Sect. 4.2, cf. Eq. (4.27), in view of the fact that the width of single-particle states depend linearly with T (cf. Eq. (9.10)). It becomes quite natural, once the interference effects between the vertex and self-energy contributions to the self-energy are taken into account. This can be seen from Fig. 9.6, where the full width at half maximum of the strength function for the giant dipole resonance of ^{120}Sn is shown as a function of T. Two calculations of the self-energy of the resonance have been carried out, one in which only the contributions of the type given in Eq. (9.18) are taken into account. That is, contributions in which the particle (hole) in the intermediate state excites and reabsorbs a surface vibration (cf. Figs. 4.12(a) and (b)). The corresponding width displays a linear dependence with T. Furthermore, it is essentially equal to twice

$$\Gamma_p^\downarrow(\omega = 8 \text{ MeV}, T) \approx \Gamma_h^\downarrow(\omega = 8 \text{ MeV}, T) \approx 4.1 \text{ MeV} + 0.2T, \quad (9.20)$$

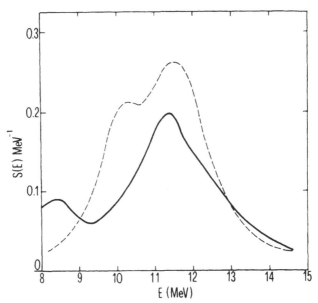

Figure 9.5. Strength function for the isoscalar giant quadrupole resonance of ^{208}Pb at $T = 0$ (dashed line) and $T = 3$ MeV (full line). (From Bortignon et al. (1986).)

in keeping with the relation given in Eq. (4.27), and with the fact that $\hbar\omega_{GDR} \approx$ 16 MeV in ^{110}Sn. The second calculation of the giant resonance strength function, whose results are also displayed in Fig. 9.6, was carried out taking into account both the contribution shown in Eq. (9.17) as well as that displayed in Eq. (9.18) (total self-energy, cf. Figs. 4.12(a–d)). The corresponding value of the FWHM is smaller by 30–40% from the previous one. Furthermore, it is almost independent of temperature. This result is at variance with the classical result obtained by Landau (1957) on the temperature dependence of the attenuation of zero sound, and underscores the role that the surface plays in the damping process of nuclear motion.

In keeping with the discussion carried out in Sect. 4.2.2, the contribution to the damping width to be added to the result given above and arising from the coupling to the compound nucleus, is controlled by the quantity $R_0 = \sqrt{n_0}v_0$. From the results of Sect. 8.4.2, it is seen that $v_0 \sim \beta_L$ decreases with temperature, a decrease which is largely compensated by the increase in n_0. Furthermore, note that W_0 will not depend sensitively on temperature. Consequently, although one cannot rule out a temperature dependence of Γ_0, this is expected to be weak and then will not change the estimates of Sect. 4.2.2 and the results given above.

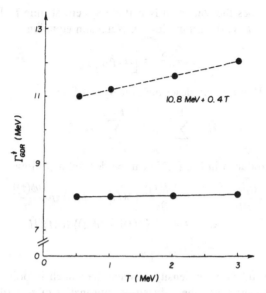

Figure 9.6.   Damping width of the GDR of ^{120}Sn as a function of temperature. The continuous line joins the results obtained taking into account all processes contributing to the giant vibration self-energy. The dashed line joins the results obtained taking into account only the contributions of the type shown in Eq. (9.18). (From Donati et al. (1996a).)

9.3   Non-Perturbative Treatment of Collisions

In the discussion of the relaxation of single-particle and of vibrational motion carried out in the present chapter ($T \neq 0$), as well as in Ch. 4 ($T = 0$), the coupling of these modes to surface fluctuations appears to play a central role. The damping of the coupling to other degrees of freedom of the nucleus, which are partially taken into account through the coupling to non-collective, multiparticle–multihole excitations of the system, seems to play a minor role. Although this scheme seems sound, it is based on two assumptions: a) the validity of perturbation theory, b) the dominance of particle–hole correlation over particle–particle collectivity.

Ultimately, the damping of nuclear motion can be traced back to collisions among the nucleons. The Time-Dependent Density-Matrix Formalism (TDDM), (Wang and Cassing (1985)) provides a non-perturbative scheme to deal with these processes. In what follows, we shall discuss this formalism and an application to the damping of giant resonances as a function of temperature.

9.3.1   Density-matrix formalism

Let us introduce the N-body density matrix defined as

$$\hat{\rho}_N(t) = |\phi(t)\rangle \langle \phi(t)|, \tag{9.21}$$

where $\phi(t)$ describes the state of a N-particle system at time t. The equation of motion for $\hat{\rho}_N(t)$ is the Liouville–von Neumann equation

$$i\hbar\frac{\partial \hat{\rho}_N}{\partial t} = \left[\hat{H}, \hat{\rho}_N\right], \tag{9.22}$$

where $\hat{H}$ is the Hamiltonian describing the system

$$\hat{H} = \sum_{i=1}^{N} h_o(i) + \sum_{i<j}^{N-1} v(i,j). \tag{9.23}$$

The relation introduced in Eq. (9.22) can be derived as follows

$$\begin{aligned}
i\hbar\frac{\partial \hat{\rho}_N}{\partial t} &= i\hbar\frac{\partial |\phi(t)\rangle}{\partial t}\langle\phi(t)| + |\phi(t)\rangle\, i\hbar\frac{\partial\langle\phi(t)|}{\partial t} \\
&= \hat{H}|\phi(t)\rangle\langle\phi(t)| - |\phi(t)\rangle\langle\phi(t)|\hat{H} \\
&= \left[\hat{H}, \hat{\rho}_N\right].
\end{aligned} \tag{9.24}$$

Eventually, the full N-body density contains too much information. It is in general useful to introduce the reduced density-matrix of n particles ($n \leq N$) as

$$\hat{\rho}_n(1,2,...,n;1',2',...,n';t) = \frac{1}{(N-n)!}Tr_{n+1,..,N}\left(\hat{\rho}_N\right). \tag{9.25}$$

The equation of motion for this quantity reads

$$\begin{aligned}
i\hbar\frac{\partial \hat{\rho}_n}{\partial t} &= \left[\sum_{i=1}^{n} h_o(i), \hat{\rho}_n\right] + \left[\sum_{i<j}^{n-1} v(i,j), \hat{\rho}_n\right] \\
&\quad + Tr_{n+1}\left(\left[\sum_{i=1}^{n} v(i,n+1), \hat{\rho}_{n+1}\right]\right),
\end{aligned} \tag{9.26}$$

while the associated expectation value of a n-body operator $\hat{A}$, is given by

$$\langle\hat{A}\rangle = \frac{1}{n}Tr(\hat{A}\hat{\rho}_n). \tag{9.27}$$

The equation of motion of the n-body density matrix involves the $n+1$-body density matrix. Thus, the relation introduced in Eq. (9.26) is equivalent to a set of coupled equations, known as the BBGKY (Born, Bogoliubov, Green, Kirkwood and Yvon) hierarchy, Bogoliubov (1968). In principle, to solve this set of coupled equations is as hard as to solve Eq. (9.22). However, it is easier to make approximations on the equations of the hierarchy than in the Liouville–von Neumann equation.

9.3.2 The time-dependent density-matrix formalism

The TDDM approach to the solution of Eq. (9.26) consists in closing the system of equations at the level of $\hat{\rho}_2$, that is by neglecting the correlations including three and more nucleons. In keeping with this approximation, one considers the equations of motion

$$i\hbar \frac{\partial \hat{\rho}_1}{\partial t} = [h_o(1), \hat{\rho}_1] + Tr\left([v, \hat{\rho}_2]\right), \tag{9.28}$$

and

$$\begin{aligned} i\hbar \frac{\partial \hat{\rho}_2}{\partial t} = & [h_o(1) + h_o(2), \hat{\rho}_2] + [v, \hat{\rho}_2] \\ & + Tr\left([v(1,3) + v(2,3), \hat{\rho}_3]\right), \end{aligned} \tag{9.29}$$

and writes the density matrices $\hat{\rho}_2, \hat{\rho}_3$ as a sum of a free and of a correlation term c_n, that is,

$$\hat{\rho}_2 = \mathcal{A}\mathcal{S}_2(\hat{\rho}_1\hat{\rho}_1) + c_2, \tag{9.30}$$

and

$$\hat{\rho}_3 = \mathcal{A}\mathcal{S}_3(\hat{\rho}_1\hat{\rho}_1\hat{\rho}_1 + \hat{\rho}_2 c_2) + c_3. \tag{9.31}$$

The operator $\mathcal{A}\mathcal{S}_n$ ensures the symmetry and antisymmetry properties of the corresponding density matrix. The TDDM approximation is obtained setting $c_3(t) = 0$.

For $c_2(t) = 0$, one obtains the Time-Dependent Hartree–Fock (TDHF) equations equivalent, in the small-amplitude vibration limit, to the RPA equations. It is then natural to expand both the one-body density matrix $\rho_1(1, 1'; t) \equiv \rho(1, 1'; t)$ and the two-body correlation function $c_2(12, 1'2'; t)$ in terms of single-particle states ψ_α fulfilling the TDHF equations

$$(i\hbar \frac{\partial}{\partial t} - h(1))\psi_\alpha(1, t) = 0, \tag{9.32}$$

according to

$$\rho(1, 1', t) = \sum_{\alpha,\beta} n_{\alpha\beta} \psi_\beta^*(1', t)\psi_\alpha(1, t), \tag{9.33}$$

and

$$c_2(1, 2, 1', 2'; t) =$$

$$= \sum_{\alpha,\beta\alpha',\beta'} C_{\alpha\beta\alpha'\beta'}(t)\psi_\alpha(1, t)\psi_\beta(2, t)\psi_{\alpha'}^*(1', t)\psi_{\beta'}^*(2', t). \tag{9.34}$$

The quantity $h(i) = t(i) + U(i)$ is the one-body Hamiltonian, that is, the sum of a kinetic energy and of a mean field term

$$U(i; t) = Tr_{2=2'}\left(v(i, 2)A_{i2}\rho(2, 2'; t)\right). \tag{9.35}$$

The quantity $v(12)$, is the two-body interaction acting among nucleons, while $A_{12} = 1 - P_{12}$, where P_{12} denotes the permutation operator between nucleons.

The equations of motion for the occupation matrix $n_{\alpha\beta}(t)$ and for the correlation coefficients $C_{\alpha\beta'\beta'}(t)$ are

$$i\hbar\frac{\partial n_{\alpha\beta}}{\partial t} = \sum_{\gamma\delta\sigma}[C_{\gamma\delta\beta\sigma}\langle\alpha\sigma|v|\gamma\delta\rangle - C_{\alpha\delta\gamma\sigma}\langle\gamma\sigma|v|\beta\delta\rangle], \tag{9.36}$$

and

$$i\hbar\frac{\partial C_{\alpha\beta\alpha'\beta'}}{\partial t} = B_{\alpha\beta\alpha'\beta'} + P_{\alpha\beta\alpha'\beta'} + H_{\alpha\beta\alpha'\beta'}. \tag{9.37}$$

The quantity

$$B_{\alpha\beta\alpha'\beta'} = \sum_{\lambda_1\lambda_2\lambda_3\lambda_4} \langle\lambda_1\lambda_2|v|\lambda_3\lambda_4\rangle_A$$

$$[(\delta_{\alpha\lambda_1} - n_{\alpha\lambda_1})(\delta_{\beta\lambda_2} - n_{\beta\lambda_2})n_{\lambda_3\alpha'}n_{\lambda_4\beta'}$$

$$- n_{\alpha\lambda_1}n_{\beta\lambda_2}(\delta_{\lambda_3\alpha'} - n_{\lambda_3\alpha'})(\delta_{\lambda_4\beta'} - n_{\lambda_4\beta'})], \tag{9.38}$$

is the lowest order contribution arising from collisions in the particle–particle channel (Born approximation), while the term P represents the higher order particle–particle (and hole–hole) contributions

$$P_{\alpha\beta\alpha'\beta'} = \sum_{\lambda_1\lambda_2\lambda_3\lambda_4} \langle\lambda_1\lambda_2|v|\lambda_3\lambda_4\rangle$$

$$[\delta_{\alpha\lambda_1}\delta_{\beta\lambda_2}C_{\lambda_3\lambda_4\alpha'\beta'} - \delta_{\lambda_3\alpha'}\delta_{\lambda_4\beta'}C_{\alpha\beta\lambda_1\lambda_2}$$

$$- \delta_{\alpha\lambda_1}n_{\beta\lambda_2}C_{\lambda_3\lambda_4\alpha'\beta'} - \delta_{\lambda_2\beta}n_{\alpha\lambda_1}C_{\lambda_4\lambda_3\beta'\alpha'}$$

$$+ \delta_{\lambda_3\alpha'}n_{\lambda_4\beta'}C_{\alpha\beta\lambda_1\lambda_2} + \delta_{\lambda_4\beta'}n_{\lambda_3\alpha'}C_{\alpha\beta\lambda_1\lambda_2}]. \tag{9.39}$$

The last term in Eq. (9.37), that is,

$$H_{\alpha\beta\alpha'\beta'} = \sum_{\lambda_1\lambda_2\lambda_3\lambda_4} \langle\lambda_1\lambda_2|v|\lambda_3\lambda_4\rangle$$

$$[\delta_{\alpha\lambda_1}(n_{\lambda_3\alpha'}C_{\beta\lambda_4\beta'\lambda_2} - n_{\lambda_3\beta'}C_{\beta\lambda_4\alpha'\lambda_2}$$

$$- n_{\lambda_4\alpha'}C_{\lambda_3\beta\lambda_2\beta'} - n_{\lambda_4\beta'}C_{\lambda_3\beta\alpha'\lambda_2})$$

$$+ \delta_{\beta\lambda_2}(n_{\lambda_4\beta'}C_{\alpha\lambda_3\alpha'\lambda_1} - n_{\lambda_4\alpha'}C_{\alpha\lambda_3\beta'\lambda_1}$$

$$- n_{\lambda_3\beta'}C_{\alpha\lambda_4\alpha'\lambda_1} - n_{\lambda_3\alpha'}C_{\alpha\lambda_4\lambda_1\beta'})$$

$$- \delta_{\beta'\lambda_4}(n_{\beta\lambda_2}C_{\alpha\lambda_3\alpha'\lambda_1} - n_{\alpha\lambda_2}C_{\beta\lambda_3\alpha'\lambda_1}$$

$$-n_{\beta\lambda_1}C_{\alpha\lambda_3\alpha'\lambda_2} - n_{\alpha\lambda_1}C_{\beta\lambda_3\lambda_2\alpha})$$

$$-\delta_{\alpha'\lambda_3}(n_{\alpha\lambda_1}C_{\beta\lambda_4\beta'\lambda_2} - n_{\alpha\lambda_2}C_{\alpha\lambda_4\beta'\lambda_2}$$

$$- n_{\alpha\lambda_2}C_{\beta\lambda_4\beta'\lambda_1} - n_{\beta\lambda_2}C_{\alpha\lambda_4\lambda_1\beta'})] \tag{9.40}$$

is the contribution to the equations of motion of collisions in the particle–hole channel (density fluctuations).

The TDDM equations reported above fulfill the conservation laws of particle number, momentum and energy. For the particle-number operator $N = \sum_\alpha n_{\alpha\alpha}$, one may write, according to Eq. (9.36),

$$i\hbar\frac{dN}{dt} = \sum_{\alpha\gamma\delta\sigma}[C_{\gamma\delta\alpha\sigma}\langle\alpha\sigma|v|\gamma\delta\rangle - C_{\alpha\delta\gamma\sigma}\langle\gamma\sigma|v|\alpha\delta\rangle]$$

$$= 0. \tag{9.41}$$

Energy conservation implies that

$$\frac{dE}{dt} = Tr\left(\hat{H}\dot{\rho}_N\right)$$

$$= Tr_1\left(t(1)\dot{\rho}_1(1)\right) + \frac{1}{2}Tr_{1,2}\left(v(1,2)\dot{\rho}_2\right)$$

$$= \frac{-i}{\hbar}Tr\left(\rho_1[t(1),t(1)]_-\right) + \frac{-i}{\hbar}Tr\left(t(1)[v(1,2),\rho_2]_-\right)$$

$$+ \frac{i}{\hbar}Tr\left(t(1)[v(1,2),\rho_2]_-\right) + \frac{-i}{\hbar}Tr\left(\rho_2[v(1,2),v(1,2)]_-\right)$$

$$- \frac{i}{\hbar}Tr\left(\rho_3[v(1,2),v(1,3)+v(2,3)]_-\right) = 0 \tag{9.42}$$

Consequently, the approximations introduced for $\hat{\rho}_3$ do not affect the conservation laws connected to one- and two-body operators.

9.3.3 Damping of giant vibrations

In what follows we discuss the results of the solution of the coupled Eqs. (9.36) and (9.37) as a function of temperature, for the isovector dipole and the isoscalar quadrupole vibrations in the nuclei ^{16}O and ^{40}Ca.

For the interaction v appearing in the mean-field potential of Eq. (9.35), a force of Skyrme-type was used while a contact interaction $v(12) = -V_0\delta(\mathbf{r}_1 - \mathbf{r}_2)$ was employed for the residual interaction appearing in Eqs. (9.36)–(9.40). In keeping with the fact that, due to Pauli blocking, the matrix elements of $v(1,2)$ at saturation density ρ_0 (≈ 0.17 fm^{-3}) are quite small as compared to the matrix elements calculated at lower densities, thus making the nuclear surface the main source of damping of nuclear motion, the strength V_0 has been determined from the strength of Skyrme-type effective forces calculated at density $\rho = \rho_0/2 \approx 0.08$ fm^{-3}. The resulting values for V_0 are in the range 300–400 MeV fm^3. The set of single-particle levels used in the calculations include the 1s, 1p, 2s and 1d orbitals in the case of ^{16}O and 1s, 1p, 2s, 1d, 2p and 1f states in the case of ^{40}Ca. To carry out calculations at finite temperature, the initial occupation numbers are defined by the appropriate Fermi distributions. The system is then propagated in time (typically a time τ of the order of $\sim 0.2 \cdot 10^{-21}$s), to build up its correlations before exciting the collective modes of interest. To do this, the correlated ground state is boosted by applying appropriate phase factors to the single-particle wave functions, that is,

$$\psi_\lambda(\vec{r}, t = \tau^+) = \exp(i\alpha F)\phi_\lambda(\vec{r}). \qquad (9.43)$$

Here $F(\vec{r})$ is a one-body operator with the symmetries of the collective mode under study, for example a dipole or a quadrupole field. In this way the system acquires a well defined collective energy E_{coll} proportional to the strength factor α^2,

$$E_{coll} = \frac{\alpha^2 \hbar^2}{2m} \int d\vec{r}\rho(\vec{r})[\vec{\nabla}F]^2, \qquad (9.44)$$

where $\rho(\vec{r})$ is the density of the nucleus in $\vec{r}$-space.

Expanding the right hand side of Eq. (9.43) up to first order in α, as in a linear response regime, one obtains

$$\psi_\lambda(\vec{r}, t = \tau^+) = (1 + i\alpha F)\phi_\lambda(\vec{r}). \qquad (9.45)$$

The above equation implies that the single-particle wavefunctions ψ are the solutions of the Schrödinger equation containing a potential

$$V(\vec{r}, t) = -\delta(t)\hbar\alpha F, \qquad (9.46)$$

which results from the impulsive action of the external field. The evolution of the collective response of the system is followed in time, both in TDHF and in TDDM, by calculating its $Q_F(t)$-moments defined as

$$\langle Q_F(t) \rangle = \int d\vec{r}\rho(\vec{r}, t)F. \qquad (9.47)$$

The Fourier transform of $\langle Q_F(T) \rangle$, that is,

$$S_F(E) = \frac{1}{\pi\hbar\alpha} \frac{1}{1 - \exp(-E/T)} \int_0^\infty dt\langle Q_F(t) \rangle \sin(Et/\hbar), \qquad (9.48)$$

gives the strength function of the system associated with the Q_F-moment. As expected, this quantity is independent of α, and is defined as

$$S_F(E) = \frac{1}{Z} \sum_{i,j} \exp(-E_i/T)\langle i|F|j\rangle^2 \delta(E - E_i + E_j), \qquad (9.49)$$

where Z is the partition function $Z = \sum_i \exp(-E_i/T)$, the sum being over all the states i of the system. From the strength function, one can calculate the contribution S_1

$$S_1 = \int dE S_F(E)E \qquad (9.50)$$

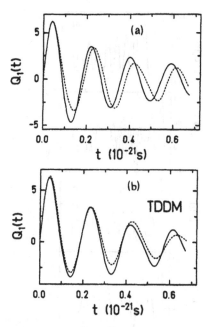

Figure 9.7. (a) The dipole moment $Q_1(t)$ of ^{40}Ca at $T = 0$ in the limits TDHF (solid line) and TDDM (dashed line). (b) The dipole moment $Q_1(t)$ in ^{40}Ca in the limit TDDM at temperature $T = 0$ (solid line) and $T = 4$ MeV (dashed line). (Adapted from De Blasio et al. (1992).)

to the Energy Weighted Sum Rule (EWSR) of the operator F

$$(\text{EWSR})_F = \frac{1}{2}\langle [F, [H, F]] \rangle. \tag{9.51}$$

In Fig. 9.7(a), the time-dependent evolution of the dipole moment of ^{40}Ca calculated at $T = 0$ in the TDHF and TDDM approximations are compared. In Fig. 9.7(b), the evolution of the same dipole operator calculated in TDDM is shown as a function of time for two temperatures, namely $T = 0$ and $T = 4$ MeV.

Two features emerge from a simple inspection of these results. First, the dipole vibration displays essentially a single frequency, which is strongly damped already in TDHF. The relaxation of the giant dipole resonance of ^{40}Ca seems thus to be controlled by the decay into single-particle motion (Landau damping) and by the coupling to the continuum (escape width). Second, the properties of the resonance are unaffected by temperature. This is a natural consequence of the fact that damping is in this case, controlled by mean-field effects.

Fitting the main frequency of the vibration in terms of a damped oscillator whose coordinate is parametrized according to

$$\langle Q_F(t) \rangle = Q_{Fo} \sin(\omega t) \exp(-\gamma t/\hbar), \tag{9.52}$$

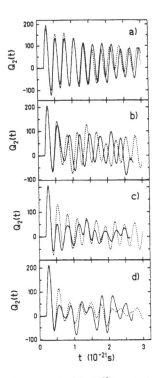

Figure 9.8. The quadrupole moment $Q_2(t)$ of ^{40}Ca calculated making the following approximations: (a) TDHF, (b) TDHF plus residual "two-body collisions," $H = 0$ in Eq. (9.37), (c) TDHF plus fluctuations, $P = 0$ in Eq. (9.37), (d) full TDDM. In all cases, the calculations were carried out for two temperatures, namely $T = 0$ (solid lines) and $T = 4$ MeV (dashed lines). (Adapted from De Blasio et al. (1992).)

one obtains a damping width $\Gamma^{\downarrow} = 2\gamma$ which is about 4 MeV and compares well with the experimental value of 5 MeV. Similar results are obtained in the case of ^{16}O, where the calculated widths of the order of 5 MeV are essentially independent of temperature, as well as of the presence of collisions.

In Fig. 9.8, the time evolution of the quadrupole moment of ^{40}Ca is shown. It was calculated at two temperatures, namely at $T = 0$ and $T = 4$ MeV and in a number of approximations: (a) TDHF, (b) TDDM, but neglecting collisions leading to particle–hole excitations, that is density fluctuations ($H_{\alpha\beta\alpha'\beta'} = 0$ in Eq. (9.37)), (c) TDDM, but neglecting collisions in the particle–particle and hole–hole channels, ($P_{\alpha\beta\alpha'\beta'} = 0$ in Eq. (9.37)), (d) TDDM including all collisional processes.

The giant quadrupole resonance (GQR) appears in TDHF (Fig. 9.8(a)) as a single mode which is only slightly damped and whose properties are independent of temperature. The presence of collisions and density fluctuations changes this picture in a qualitative way by producing a strongly damped vibration displaying a complicated beating pattern, pointing to the existence of a variety of normal modes. These effects are mainly controlled by fluctuations (collisions leading to particle–hole excitations cf. Fig. 9.8(c)) while collisions in the particle–particle and hole–hole channels (Fig. 9.8(b)) play a minor role. In any case, the role of these channels cannot be neglected in a quantitative description of the damping processes, because of the interference with the particle–hole channel (cf. Fig. 9.8(d)). Applying the fitting procedure introduced in Eq. (9.52) to the main peak of the time-dependent quadrupole vibration of ^{40}Ca, one obtains values for the damping width of the order of 5 MeV at $T = 0$ and 3 MeV at $T = 4$ MeV. The collective quadrupole response of ^{16}O and ^{32}S were also studied and the results display similar features to those obtained for ^{40}Ca.

Time Dependent Density Matrix Theory treats the main relaxation mechanisms found in many-body systems, namely: escape of particles in the continuum, Landau damping, and collisional damping associated with both particle collisions and with density fluctuations, on essentially equal footing. It is found that three of these mechanisms are, in the nuclear case, independent of temperature. The first two, in keeping with the fact that they are mean-field effects. The last because collisions, already at $T = 0$ MeV, lead to a modification of the single-particle occupation factors around the Fermi energy similar to that produced by a temperature of a few MeV in a system of non-interacting fermions (cf. Fig. 4.9 and Eq. (7.19)). Temperatures of the order of few MeV hardly affect such a picture. The effect of the only mechanism (collisional damping in the p–p and h–h channels) that depends on temperature is too weak to be observed, at least within the range of temperatures under discussion (0 MeV $\leq$ T $\leq$ 4 MeV).

As already mentioned in Sect. 9.1.1, this result is very different from the infinite system result. In fact, the absorption coefficient of zero sound of such a system can be written as (Landau (1957))

$$\gamma \propto (\hbar\omega)^2 [1 + (2\pi T/\hbar\omega)^2]. \tag{9.53}$$

The T^2 dependence can also be obtained in the TDDM approach, by retaining only the B term in Eq. (9.37). It is essentially produced by the occupation factors. The B term corresponds to the collision integral of the nuclear transport-equation approaches, but its contribution to the damping observed in the results displayed in Figs. 9.7 and 9.8 is rather small (10%), and is masked by the much larger contributions of the P and H terms.

10

Dipole Oscillations: Theory

At finite temperature, the nucleus can be in any of the possible mean field configurations compatible with the associated thermal ensemble. Around each of these configurations it displays quantal fluctuations. In what follows we shall use the theoretical tools developed in previous chapters to try to disentangle the role played by static deformations, and by quantal and large amplitude thermal fluctuations, in determining the observed properties of the giant dipole resonance.

10.1 Effective Charges and Centre-of-Mass Correction

The electric-dipole interaction of the spatially constant electromagnetic field with the nucleons can be written as

$$
\begin{aligned}
H & = e\mathcal{E} \sum_p z_p \\
& = e\mathcal{E} \left[\frac{N}{A} \sum_p z_p + \frac{Z}{A} \sum_p z_p + \frac{Z}{A} \sum_n z_n - \frac{Z}{A} \sum_n z_n \right] \\
& = e\mathcal{E} \left[\frac{N}{A} \sum_p z_p - \frac{Z}{A} \sum_n z_n + Z\bar{z} \right],
\end{aligned} \tag{10.1}
$$

where p stands for protons, n for neutrons, and

$$
\bar{z} = \frac{\sum_i z_i}{A}, \tag{10.2}
$$

denotes the position of the centre of mass of the nucleus. The term $e\mathcal{E} Z\bar{z}$ corresponds to the interaction of the entire nucleus with the electric field and leads to the nuclear Thomson scattering. The term $e\mathcal{E}[\frac{N}{A} \sum_p z_p - \frac{Z}{A} \sum_n z_n]$ gives the dipole E1-photon absorption due to internal motion inside the nucleus. That is, each proton acts as if it had an "effective charge" $e\frac{N}{A}$, and each neutron as if it had an "effective charge" $-e\frac{Z}{A}$ (cf. Levinger (1960)).

10.2   The Oscillator Strength

If a quantum-mechanical system, originally in its ground state $|0\rangle$, is perturbed for a time t, then the probability $|a_n|^2$ of finding the system in a discrete state n is given by time-dependent perturbation theory as

$$|a_n(t)|^2 = |-\frac{i}{\hbar}\int_0^t \exp[i\omega_{0n}t']H(\vec{r},t')_{0n}dt'|^2.   \qquad (10.3)$$

Here H is the perturbation acting on the nucleus, the subscripts 0 and n denote the matrix element between stationary states, and the frequency

$$\omega_{0n} = \frac{E_n - E_0}{\hbar}.   \qquad (10.4)$$

In the electric-dipole approximation, the perturbation H is due to a spatially constant electric field along the z-axis of amplitude $\mathcal{E}$ and angular velocity ω. We shall consider the perturbation due to the electric field interacting with a single-particle of charge e to be

$$H(r,t)_{0n} = -e(z)_{0n}\mathcal{E}\cos\omega t.   \qquad (10.5)$$

Substituting in Eq. (10.3) and carrying out the integration over dt', one finds

$$|a_n(t)|^2 = e^2\mathcal{E}^2(z)_{0n}^2\left[\frac{\sin^2(\omega-\omega_{0n})\frac{t}{2}}{\hbar^2(\omega-\omega_{0n})^2} + \frac{\sin^2(\omega+\omega_{0n})\frac{t}{2}}{\hbar^2(\omega+\omega_{0n})^2}\right].   \qquad (10.6)$$

This function displays a sharp resonance at $\omega_D = \omega_{0n} = \frac{E_n - E_0}{\hbar}$. In order to obtain an expression for the probability that varies monotonically with t we integrate over ω for a small range of frequencies around ω_n and obtain

$$P = \int_{\text{line}}|a_n(t)|^2 d\omega = e^2\mathcal{E}^2(z)_{0n}^2\frac{\pi t}{2\hbar^2}.   \qquad (10.7)$$

As expected, the transition rate $\frac{dP}{dt}$ is independent of t. The cross-section σ for photon absorption is defined as

$$\int_{\text{line}}\sigma_{0n}d\omega = \frac{\int_{\text{line}}(\text{transitions/s})d\omega}{\text{photon flux}} = \frac{4\pi^2 e^2 W(z)_{0n}^2}{\hbar^2 c}(1+K),   \qquad (10.8)$$

where we have used the photon energy $W = \hbar\omega_D$ and photon flux $= \frac{e\mathcal{E}^2}{8\pi\hbar\omega_D}$. The integral defined in Eq. (10.8) is proportional to the dipole energy weighted sum rule, while the factor $(1+K)$ takes care of the energy dependent part of the single-particle potential (cf. Eq. (3.37) and subsequent discussion). This effect is considered through integration over the experimental line shape. The empirical value of K depends on the range of energy over which the photoabsorption cross-section is integrated. Performing the integration up to twice the giant dipole energy centroid, values of K of the order of 0.2 are obtained. Typical results obtained making use of Skyrme interaction are about twice this value (Lipparini and Stringari (1989)).

The oscillator strength for E1-transitions between discrete states is defined as

$$f_{0n} = \frac{2mW}{\hbar^2}(z)_{0n}^2 = (\frac{z_{0n}}{\lambda_{0n}})^2, \tag{10.9}$$

where $\lambda_{0n} = \frac{1}{k_{0n}}$ is the de Broglie wavelength for a particle of energy $\hbar\omega_D$, i.e., $\lambda_{0n} = \frac{\hbar}{p_{0n}}$. Equation (10.8) is now rewritten as

$$\int_{\text{line}} \sigma_{0n}dW = \frac{2\pi^2 e^2 \hbar}{mc} f_{0n}. \tag{10.10}$$

As seen from Eq. (10.9), f_{0n} is a dimensionless quantity. In what follows we show that

$$\sum_n f_{0n} = 1, \tag{10.11}$$

which is another expression of the Thomas–Reiche–Kuhn sum rule (TRK) (cf. Eq. (3.40)).

Relation (10.11) is derived using Heisenberg matrix relations

$$W z_{0n} = (E_n - E_0)z_{0n} = -[H, z]_{0n} = [z, H]_{0n}, \tag{10.12}$$

where H is the Hamiltonian operator and the square brackets denote the commutator. Taking care to keep our quantities Hermitian by using the Heisenberg relations alternatively for the first and second z_{0n} in Eq. (10.9), we find that the summed oscillator strength is given by

$$
\begin{aligned}
\sum_n f_{0n} &= \frac{2m}{\hbar^2}\sum_n (E_n - E_0)z_{0n}z_{0n} \\
&= -\frac{m}{\hbar^2}\sum_n ([H, z]_{0n}z_{0n} + z_{0n}[H, z]_{0n}). \tag{10.13}
\end{aligned}
$$

Using closure, we express the summed oscillator strength in terms of the properties of the ground state alone:

$$\sum_n f_{0n} = -\frac{m}{\hbar^2}([H, z]z - z[H, z])_{00} = -\frac{m}{\hbar^2}([[H, z], z])_{00}, \tag{10.14}$$

where we have used $z_{0n} = z_{n0}$ and $[H, z]_{0n} = -[H, z]_{n0}$. Substituting $H = \frac{p^2}{2m} + V(r)$, we find

$$\sum_n f_{0n} = -\frac{m}{\hbar^2}([[\frac{p^2}{2m}, z], z])_{00} - \frac{m}{\hbar^2}([[V, z], z])_{00}. \tag{10.15}$$

The second term is zero if the potential V is a function only of position. Using $[p, z] = -i\hbar$, one can show that the first term of Eq. (10.15) is unity.

The quantum-mechanical TRK sum rule given by Eq. (10.11) corresponds to the classical integrated power absorption for forced oscillators by a charged oscillator. From Eq. (10.7) and the TRK sum rule one obtains the power absorption integrated over all frequencies for a charged oscillator to be

$$\frac{W}{t} \sum_n \int |a_n(t)|^2 d\omega = \frac{\pi}{4} \frac{e^2 \mathcal{E}^2}{m} \tag{10.16}$$

for any form $V(r)$ of the oscillator potential. This quantum-mechanical result is in exact agreement with the classical treatment.

10.3 Polarizability Sum Rule

A second classical interpretation of the oscillator strength is obtained by comparing the results of the quantum-mechanical and classical calculations of the electric polarizability. From a quantum-mechanical second-order time-independent perturbation theory we have

$$\alpha = \frac{\text{dipole moment}}{\text{electric field}} = 2e^2 \sum_n \frac{(z)_{0n}^2}{E_n - E_0}. \tag{10.17}$$

Expressing $(z)_{0n}^2$ in terms of the oscillator strength f_{0n} defined in Eq. (10.9), one obtains

$$\alpha = e^2 \sum_n \frac{f_{0n}}{m\omega_{0n}^2} = e^2 \sum_n \frac{f_{0n}}{k_{0n}}, \tag{10.18}$$

with the spring constant $k_{0n} = m\omega_{0n}^2$. This quantum-mechanical result agrees with a classical calculation of α if one interprets the oscillator strength f_{0n} as the fraction of the charged particles bound by a linear spring of spring constant k_{0n}.

For the case of internal nuclear excitation, the TRK sum rule is modified to give the summed internal oscillator strength

$$\sum_n f_{0n} = \left(\frac{N}{A}\right)^2 \sum_{p,p'} \delta(p,p') + \left(-\frac{Z}{A}\right)^2 \sum_{n,n'} \delta(n,n')$$

$$= \frac{N^2}{A^2} Z + \frac{Z^2}{A^2} N = \frac{NZ}{A}. \tag{10.19}$$

The oscillator strength $\frac{Z^2}{A}$ for nuclear Thomson scattering combines with $\frac{NZ}{A}$ to give the TRK value of Z.

Making use of Eqs. (10.8) and (10.19), the total photoabsorption cross-section integrated over frequency is given by

$$\int_{\text{line}} \sigma dE = \sum_n \int \sigma_{0n} dE = \frac{2\pi^2 e^2 \hbar}{mc} \frac{NZ}{A} (1 + K). \tag{10.20}$$

Following steps similar to those needed to obtain this result, one finds

$$\int \left(\frac{\sigma}{E^2}\right) dE = \frac{2\pi^2}{\hbar c} \alpha, \tag{10.21}$$

which is known as the polarizability or Migdal's sum rule (Migdal (1967)). The relation between the nuclear polarizability α and the inverse square energy is a natural consequence of Eq. (10.18).

To determine the relationship between the polarizability and the forces and dimensions of the nucleus, we examine the distortion in the nuclear charge density caused by a polarizing field. The energy of the system, in the presence of an external field, is the sum of two terms. The first is the field energy, which one writes in terms of the electric field, taken to be in the z-direction, and the deviation $\delta\varrho$ of the proton density from equilibrium,

$$E_1 = \frac{1}{2} e \mathcal{E} \int z \delta \varrho d^3 r, \qquad (10.22)$$

where the effective charge $\frac{e}{2}$ has been used ($\frac{Z}{A} \sim \frac{N}{A} \sim \frac{1}{2}$). The second contribution to the energy is the internal energy of the nucleus. It must vary quadratically with the deviation $\delta\varrho$ of the charge density from equilibrium, because the initial state is stationary. Its structure can be obtained from the symmetry term of the mass formula

$$E_2 = -\frac{1}{2\varrho_0} b_{sym} \int (\delta\varrho(r))^2 d^3 r, \qquad (10.23)$$

where b_{sym} is the nuclear symmetry coefficient and ϱ_0 is the nuclear density. We next minimize the total energy

$$E = E_1 + E_2, \qquad (10.24)$$

with respect to possible choices of the deviation $\delta\varrho$ of the charge density from equilibrium. The relation $\frac{\partial E}{\partial (\delta\varrho)} = 0$ leads to

$$\int \left(\delta\varrho - \frac{\mathcal{E} e \varrho_0}{2 b_{sym}} z \right) d^3 r = 0, \qquad (10.25)$$

and thus

$$\delta\varrho = \frac{\mathcal{E} e \varrho_0}{2 b_{sym}} z. \qquad (10.26)$$

The polarizability α defined in Eq. (10.17) can now be written as

$$\alpha = \frac{(\frac{e}{2}) \int z \delta \varrho d^3 r}{\mathcal{E}} = \frac{e^2 \varrho_0}{4 b_{sym}} \int z^2 d^3 r. \qquad (10.27)$$

Making use of the relation

$$z = \sqrt{\frac{4\pi}{3}} r Y_{10}(\hat{r}), \qquad (10.28)$$

one obtains

$$\int z^2 d^3 r = \frac{4\pi}{3} \int d\Omega |Y_{10}|^2 \int r^4 dr = \frac{4\pi}{3} \frac{R^5}{5} \sim 1.73 A R^2, \qquad (10.29)$$

where $R \sim 1.2A^{\frac{1}{3}}$ fm is the nuclear radius and A the mass number. Equation (10.27) then becomes

$$\alpha = \frac{e^2 \langle r^2 \rangle A}{12 b_{sym}}. \tag{10.30}$$

One may then determine the frequency of the dipole vibration by using the sum rules (10.20) and (10.21) in the ratio

$$(\hbar \omega_D)^2 = \frac{\int \sigma dE}{\int (\frac{\sigma}{E^2}) dE} = \frac{3\hbar^2 b_{sym}}{m \langle r^2 \rangle}(1 + K), \tag{10.31}$$

as well as Eq. (3.44).

Taking into account finite-size effects within the liquid-drop model, one obtains (Lipparini and Stringari (1989))

$$b_{sym} = \frac{b_v}{1 + \frac{5}{3}\frac{b_v}{b_s}A^{-1/3}}, \tag{10.32}$$

where $b_v = 65$ MeV is the volume symmetry coefficient, while $b_s/b_v = 2.2$ is the ratio between the surface and volume symmetry coefficients. Consequently,

$$(\hbar \omega_D)^2 = \frac{3\hbar^2 b_v}{m \langle r^2 \rangle (1 + \frac{5}{3}\frac{b_v}{b_s}A^{-1/3})}(1 + K), \tag{10.33}$$

which, for medium-heavy nuclei leads to

$$\hbar \omega_D \approx \frac{80}{A^{1/3}} \tag{10.34}$$

in agreement with the experimental findings (cf. Eq. (2.1)).

10.4 Temperature Dependence of the Parameters

In this section we shall discuss the temperature dependence of the parameters which define the main properties of the giant dipole resonance in Eq. (10.33).

10.4.1 Mean square radius

We start by calculating the temperature dependence of the mean square radius. We shall first consider the corrections to this quantity due to surface fluctuations (Esbensen and Bertsch (1983), Barranco and Broglia (1985), Barranco and Broglia (1987)), and then estimate their temperature dependence. For this purpose, we use the collective description of shape deformations, which parametrizes the nuclear radius according to the relation given in Eq. (3.14). The coefficient α_{00} (breathing mode) is fixed by volume conservation, that is,

$$\int_0^R d^3r = \frac{4\pi}{3}R^3 \approx \frac{4\pi}{3}R_0^3 + R_0^3(\sqrt{4\pi}\alpha_{00} + \sum_{\lambda \geq 2, \mu}|\alpha_{\lambda,\mu}|^2) + \mathcal{O}(\alpha^3). \tag{10.35}$$

Because we consider small (quantal)-amplitude fluctuations ($\alpha^2 \ll \alpha$), we only retain terms quadratic in the deformation parameters. Equation (10.35) leads to

$$\alpha_{00} = -\frac{1}{4\pi} \sum_{\lambda \geq 2, \mu} |\alpha_{\lambda,\mu}|^2. \tag{10.36}$$

The nuclear radius can then be written as

$$R = R_0(1 + S), \tag{10.37}$$

where

$$S = \frac{s_0 + s}{R_0}, \tag{10.38}$$

and

$$\begin{cases} s_0 = & R_0\alpha_{00}, \\ s = & R_0 \sum_{\lambda \geq 2, \mu} \alpha_{\lambda,\mu} Y^*_{\lambda,\mu}(\hat{r}). \end{cases} \tag{10.39}$$

The average value of the multipole moments of the radius in the ground state was defined as in Eq. (3.44). We now proceed to the evaluation of this quantity, assuming a nucleus with a constant density

$$\begin{aligned} \int \varrho(r)r^2 d^3r &= \varrho_0 \int d\Omega \int_0^{R_0(1+S)} r^{\lambda+2} dr \\ &= \varrho_0 \int d\Omega \frac{(R_0(1 + S))^{\lambda+3}}{\lambda + 3}. \end{aligned}$$

Expanding the integrand to lowest order in the deformation parameters, one obtains

$$(R_0(1 + S))^{\lambda+3} \approx R_0^{\lambda+3}\left(1 + (\lambda + 3)S + \frac{(\lambda + 3)(\lambda + 2)}{2}S^2\right).$$

Making use of the fact that

$$\langle 0|s|0 \rangle = 0,$$

and

$$\langle 0|s^2|0 \rangle = \sigma^2(0),$$

one obtains

$$\int \varrho(r)r^\lambda d^3r = \varrho_0 \frac{4\pi}{\lambda + 3}R_0^{\lambda+3}\left[1 + \frac{\lambda(\lambda + 3)}{2}\left(\frac{\sigma(0)}{R}\right)^2\right],$$

which together with

$$\int \varrho(r)d^3r = \frac{4\pi}{3}\varrho_0 R_0^3,$$

leads to

$$\langle r^\lambda \rangle = \frac{3}{\lambda + 3} R_0^\lambda \left[1 + \frac{\lambda(\lambda + 3)}{2} \left(\frac{\sigma(0)}{R_0} \right)^2 \right].$$

The zero-point fluctuations of the surface depend on temperature through the bosonic occupation numbers, in keeping with the assumption that the surface vibrations are harmonic. Thus

$$\sigma^2(T) = R_0^2 \sum_{n,\lambda} \frac{2\lambda + 1}{4\pi} \frac{\hbar\omega_{\lambda(n)}}{2C_{\lambda(n)}} (1 + 2n_B(n, \lambda)), \qquad (10.40)$$

where

$$n_B(n, \lambda) = \frac{1}{\exp[\hbar\omega_\lambda(n)/T] - 1},$$

is a boson occupation number. Note that these fluctuations already incorporate those contained in the shell model which are already implicitly contained in the value of the nuclear radius R_0. To avoid overcounting, we have to use instead of (10.40) the corrected quantity

$$\Delta\sigma^2(T) = \sigma^2(T) - R_0^2 \sum_\lambda \frac{2\lambda + 1}{4\pi} \frac{\varepsilon_{ph}}{2C_\lambda(ph)} (1 + 2n_B(\varepsilon_{ph})), \qquad (10.41)$$

where the parameters of the last term are determined from uncorrelated particle-hole excitations. From numerical calculations one obtains (Pacheco (1989), Bracco et al. (1992))

$$\Delta\sigma^2(T) \approx \Delta\sigma^2(0)(1 + 20(\frac{T}{\varepsilon_F})^2). \qquad (10.42)$$

To these corrections one has to add those arising from the Hartree–Fock field. Temperature-dependent Hartree–Fock calculations (cf. Fig. 8.3) using Skyrme interactions lead to

$$\langle r^2 \rangle_T^{(0)} = (\langle r^2 \rangle_T)_{HF} = \langle r^2 \rangle_{T=0}^{(0)} \left(1 + 2 \left(\frac{T}{\varepsilon_F} \right)^2 \right). \qquad (10.43)$$

Combining this result with (10.42), one obtains

$$\langle r^2 \rangle_T = \langle r^2 \rangle_{T=0} \left[1 + (\alpha + \beta)(\frac{T}{\varepsilon_F})^2 \right], \qquad (10.44)$$

where $\alpha \sim 100(\Delta\sigma(0)/R)^2$ is the contribution due to zero-point fluctuations of the surface and β stands for the thermal expansion of the Hartree–Fock field. Typical values of the parameters are $\alpha \approx \beta \approx 2$ leading to

$$\langle r^2 \rangle_T = \langle r^2 \rangle_{T=0} \left[1 + 4(\frac{T}{\varepsilon_F})^2 \right]. \qquad (10.45)$$

Thus, as expected, the nucleus expands as it is heated.

10.4.2 Symmetry coefficient

Within the liquid-drop model, the volume and surface symmetry coefficients display a quadratic dependence with temperature, given by the expressions (Lipparini and Stringari (1989))

$$b_v(T) = b_v \left(1 - 1.3 \left(\frac{T}{\varepsilon_F} \right)^2 \right), \qquad (10.46)$$

and

$$b_s(T) = b_s \left(1 - 16.6 \left(\frac{T}{\varepsilon_F} \right)^2 \right). \qquad (10.47)$$

The volume symmetry coefficient receives contributions from both kinetic and potential terms (cf. Sect. 3.2.3)

$$b_v = (b_v)_{kin} + (b_v)_{pot}. \qquad (10.48)$$

In the Fermi gas model

$$(b_v)_{kin} = \frac{2}{3} \varepsilon_F = \frac{2}{3} \frac{\hbar^2 k_F^2}{2m^*} \qquad (10.49)$$

where

$$k_F = \left(\frac{3\pi^2}{2} \rho \right)^{1/3}, \qquad (10.50)$$

is the Fermi momentum, and $m^* = m_\omega m_k / m$ is the effective nucleon mass, product of the k- and of the ω-mass (cf. Sects. 3.1.1 and 4.1.2). While the k-mass is essentially independent of temperature (cf. Fig. 8.4), being a mean-field property, the ω-mass displays a marked temperature dependence (cf. Sect. 9.1.2). To simplify the discussion we shall, in what follows, use a linear parametrization of the result given in Eq. (9.14)–(9.15), namely,

$$\frac{m^*}{m} = \left(1 - 2.5 \left(\frac{T}{\varepsilon_F} \right) \right) \qquad (10.51)$$

which is a sensible approximation to the correct result within the temperature range $0 \leq T \leq 5$ MeV. The linear dependence on T displayed by m^* underscores the fact that the mechanism which is at the basis of the nucleon ω-mass goes beyond mean-field.

The Fermi energy does not only change with temperature because of the change in the nucleon mass, but also because of the variation of the nuclear density (cf. Eq. (10.50)). Making use of the relation given in Eqs. (10.45) one can write

$$\rho(T) \approx \rho \left(1 - 6 \left(\frac{T}{\varepsilon_F} \right)^2 \right). \qquad (10.52)$$

Collecting the different contributions to the thermal dependence of $b_v(T)$, one obtains

$$b_v(T) \approx b_v(1 + 0.9(\frac{T}{\varepsilon_F}) - 2.7(\frac{T}{\varepsilon_F})^2). \qquad (10.53)$$

Consequently

$$\frac{b_s(T)}{b_v(T)} \approx \frac{b_s}{b_v} \left(1 - 0.9 \left(\frac{T}{\varepsilon_F}\right) - 13.9 \left(\frac{T}{\varepsilon_F}\right)^2\right). \qquad (10.54)$$

10.5 Energy Centroid

Making use of Eqs. (10.45), (10.51) and (10.53)–(10.54), and of the relation (cf. Lipparini and Stringari (1989))

$$1 + K(T) = (1 + K) \left(1 - 1.3 \left(\frac{T}{\varepsilon_F}\right)^2\right), \qquad (10.55)$$

one obtains

$$\hbar\omega_D(T) \approx \hbar\omega_D \left(1 + 0.7 \left(\frac{T}{\varepsilon_F}\right) - 7 \left(\frac{T}{\varepsilon_F}\right)^2\right). \qquad (10.56)$$

In this expression, the T-dependence arising from mean field effects opposes that associated with the effective nucleon mass, resulting in a weak dependence of the energy centroid of the giant dipole resonance with T. In fact, within the range $0 \leq T \leq 5$ MeV, Eq. (10.56) can be written as

$$\hbar\omega_D(T) \approx (80^{+1.6}_{-3.2})A^{-1/3} \text{ MeV}. \qquad (10.57)$$

The upper limit of the above expression is obtained for $T \approx 2$ MeV, where effective mass effects are dominant, the lower limit being reached for $T \approx 5$ MeV, where mean-field effects lead to the main contributions. In any case, the corresponding shifts are small.

10.6 The Damping Width

In what follows we shall calculate the variety of contributions to the damping width of the giant dipole resonance, associated with the processes where the energy of the resonance does not leave the system but is redistributed among all degrees of freedom, in terms of simple models which allows to concentrate on the physics of the process and in the order of magnitude of the contribution. We shall leave out from this discussion the phenomenon of Landau damping (cf. Sect. 1.4), of little relevance concerning the damping of the GDR in medium and heavy nuclei (cf. Sect. 3.3). The discussion of detailed numerical calculations of the GDR damping width is taken up in Sect. 10.8.

10.6.1 Simple estimates

Typical values of the centroid and of the width of the giant dipole resonances are 15 MeV and 5 MeV, respectively. Consequently, the nucleus oscillates through about three periods before its motion loses coherence.

The frequency of the giant dipole vibration depends on the dimensions of the nucleus (cf. Eq. (3.50)). Fluctuations of the surface will thus give rise to a frequency distribution, as schematically shown in Fig. 1.8. Particularly efficient are those of quadrupole type, because of angular-momentum coupling. A rough estimate of the associated width Γ_Q can be obtained by calculating the energy distribution of the GDR based on a ground state displaying the zero-point fluctuations associated with quadrupole harmonic motion (Bohr and Mottelson (1975)). One obtains

$$P(E) = \frac{1}{3}\sum_{k=1}^{3}\int |\Phi_0(\beta)|^2 \delta(E - \omega_k(\beta,\gamma)) d\tau \sim \exp[-\frac{(E - \hbar\omega_D)^2}{2\sigma_Q^2}],$$

$$(10.58)$$

where

$$d\tau = \beta^4 d\beta |\sin 3\gamma| d\gamma \sin\theta d\theta d\phi d\psi, \qquad (10.59)$$

is the volume element and where

$$\omega_k(\beta,\gamma) = \hbar\omega_D \exp\left[-\sqrt{\frac{5}{4\pi}}\beta\cos(\gamma + \frac{2k\pi}{3})\right], \ (k = 1, 2, 3) \qquad (10.60)$$

while

$$\sigma_Q \sim 0.3\beta\hbar\omega_D. \qquad (10.61)$$

In the adiabatic limit, the width of the GDR resulting from the coupling to the surface is then

$$\Gamma_Q \sim 2.4\sigma_Q, \qquad (10.62)$$

which for typical values of β (~ 0.3) and $\hbar\omega_D$ (~ 15 MeV) leads to widths of the order or 3 MeV, in overall agreement with the results obtained in Chapter 4. The subindex Q stands for quantal, in keeping with the fact that this width arises from the coupling of the GDR to quantal fluctuations of the surface (doorway coupling, cf. Fig. 1.8 and Sect. 4.2). This contribution to the total damping width does not vary with temperature (cf. Sects. 9.2 and 9.3, cf. also Eq. (8.37) and Fig. 8.15)).

In hot nuclei, aside from small amplitude fluctuations, the system displays also thermal large-amplitude fluctuations, which sample the whole (β, γ)-plane (cf. Fig. 1.9 and 10.1) with probabilities that are determined by the Boltzmann factors (Gallardo et al. (1985), Pacheco and Broglia (1988), Alhassid et al. (1988), Alhassid and Bush (1990)).

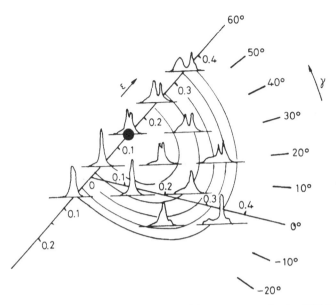

Figure 10.1. Photoabsorption cross-section in the laboratory system for the GDR in ^{108}Sn for $T = 1.5$ MeV and $I = 40\,\hbar$. The potential-energy surfaces as functions of the deformation parameters are shown as thin lines (free energy) in steps of 0.5 MeV away from the minimum, denoted by a full circle. Examples of absorption strength functions are displayed for different values of the deformation. (From Gallardo et al. (1985).)

We now proceed to calculate the contribution to the GDR damping width arising from thermal, large-amplitude fluctuations. We approximate the problem with that of a giant dipole resonance built on an ensemble of shapes of a rotating liquid drop. Within the adiabatic approximation already used in connection with Eq. (10.58), and replacing the quantal probabilities with thermal probabilities in terms of the partition function, one can calculate the associated energy dispersion. In this case, it is important to work with good angular momentum. A variety of approximations are available to carry out the corresponding projection procedure. The main effect of such a projection on the nuclear partition function, is the appearance of an effective volume element that is determined by the moments of inertia of the system (cf. Ormand et al. (1997b) and refs. therein). In the limit of a rigid-body moment of inertia, the effective volume element is found to differ only slightly from that defined in Eq. (10.59). On the other hand in the limit of irrotational flow used below, the leading behaviour in the β degree of freedom is

$$d\tau \approx \beta d\beta d\gamma. \tag{10.63}$$

In this case one obtains

$$(\Delta\omega)^2 = \frac{1}{Z} \int \beta d\beta d\gamma (\omega - \omega_k)^2 \exp[-F(\beta, \gamma)/T], \tag{10.64}$$

where

$$Z = \int \beta d\beta d\gamma \exp[-F(\beta, \gamma)/T].$$

Assuming a quadratic dependence for the free energy

$$F \approx C\beta^2,$$

and using the definition given in Eq. (10.60) for ω_k, one can carry out the integration analytically obtaining

$$\Delta\omega \approx 1.3\sqrt{T}, \qquad (10.65)$$

where the liquid-drop estimate $C \approx 70$ MeV (cf. Bohr and Mottelson (1975)) for the restoring force of the quadrupole mode in the Sn region has been used. This result implies a thermal width

$$\Gamma_T \approx 2.4\Delta\omega \approx 3\sqrt{T}\,\text{MeV}. \qquad (10.66)$$

The centrifugal force associated with the rotation of the system as a whole induces deformations, leading to a splitting of the components of the GDR along the symmetry axis and in a plane perpendicular to it. In the liquid-drop model one obtains (Bohr and Mottelson (1975))

$$\omega_0 - \omega_{\pm 1} = \frac{3}{2}\sqrt{\frac{5}{4\pi}}\beta\hbar\omega_D, \qquad (10.67)$$

where the deformation parameter

$$\beta = \sqrt{\frac{5}{4\pi}\frac{y}{1-x}}, \qquad (10.68)$$

is related to the angular momentum I through

$$y = 2.1A^{-\frac{7}{3}}I^2, \qquad (10.69)$$

and to the fissility parameter x according to

$$x = 0.0205\frac{Z^2}{A}. \qquad (10.70)$$

Assuming a Gaussian distribution, the splitting (10.66) contributes to the total width a quantity

$$\Gamma_{0\pm 1} \sim 2.4(\omega_0 - \omega_{\pm 1}). \qquad (10.71)$$

We have now all the ingredients to carry out a simple estimate of the total damping width. We do this for the case of ^{108}Sn where systematic experimental information is available (cf. Fig. 10.2). We consider the situation for which the excitation energy E^* of the compound nucleus is 51 MeV and the maximum angular momentum $I_M \sim 30\hbar$. Subtracting from E^* the rotational energy (cf. Sect. 5.4)

$$E_{rot} \sim \frac{\hbar^2}{2\mathcal{J}}I^2 \sim \frac{1}{140}30^2 \text{ MeV} \approx 6.4 \text{ MeV}, \qquad (10.72)$$

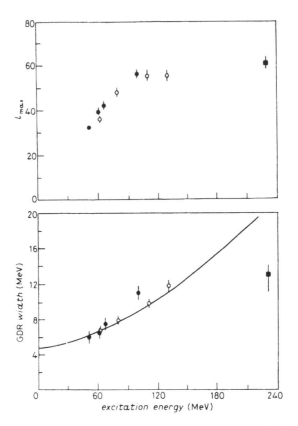

Figure 10.2. Systematic of the width of the GDR in Sn isotopes as a function of excitation energy: filled circles (Gaardhøje et al. (1986)), open circles (Chakrabarty et al. (1987)) and filled squares (Bracco et al. (1989a)). The solid line is a phenomenological fit to the data (Chakrabarty et al. (1987)), up to $E^* = 130$ MeV. Upper part: the maximum angular momentum of the compound nucleus based on the work of Grodzins et al. (1984).

one can determine the temperature of the system through the relation (cf. Eq. (5.15))

$$T = \sqrt{\frac{E^* - E_{rot} - \hbar\omega_D}{a}} \approx 1.6 \text{ MeV}, \qquad (10.73)$$

where we have used the level density parameter $a \sim \frac{A}{10}$ MeV^{-1} ≈ 10.8 MeV^{-1} and the energy centroid $\hbar\omega_D \approx 17$ MeV. One thus obtains

$$\begin{cases} \Gamma_Q & \sim 4 \text{ MeV}, \\ \Gamma_T & \sim 3\sqrt{T} \sim 4 \text{ MeV}, \\ \Gamma_{0,\pm 1} & \sim 0.5 \cdot 10^{-2}(\frac{2}{3}I_M)^2 \sim 2 \text{ MeV}. \end{cases} \qquad (10.74)$$

In the estimate of the last quantity the average value of the angular momentum was used. Considering all these contributions as arising from independent processes, one obtains

$$\Gamma_{FWHM} \sim (\sum_i \Gamma_i^2)^{\frac{1}{2}} \sim 6 \text{ MeV}, \qquad (10.75)$$

a result which is in overall agreement with the experimental findings (cf. Fig. 10.2).

10.7 Motional Narrowing

In the description of the large-amplitude thermal fluctuations carried above (cf. also Sect. 7.2.2), the ansatz of adiabaticity plays a central role. That is, it is assumed that the giant dipole vibration experiences the different (β, γ)-configurations sampled by the nucleus as static deformations. The corrections to this ansatz must depend on the time it takes for a given configuration displaying fixed values of β and γ, and also of the orientation Ω, to relax into the compound nucleus.

In what follows, we discuss a theoretical model which allows for departures from adiabaticity (Lauritzen et al. (1988), Broglia et al. (1988), Broglia (1988), Alhassid and Bush (1989), (1990a), cf. also Sects. 10.8.3 and 11.5). In accordance with the compound nucleus hypothesis, it is assumed that the ensemble of deformations at a given spin I is contained in each energy eigenstate

$$|\alpha(I)\rangle = \sum_\mu X_\mu^\alpha |\mu(\beta, \gamma, \Omega; I)\rangle, \qquad (10.76)$$

which we write as a linear combination of multi-quasiparticles $|\mu\rangle$, the average number of which is fixed by T. Each multiquasiparticle configuration has a certain intrinsic equilibrium deformation and orientation with respect to the laboratory coordinate indicated by the deformation parameters (β, γ) and by the Euler angle Ω. This is because by exciting quasiparticles one changes the number of nodes of the wavefunction describing the motion of the quasiparticles and thus the deformation of the system. The Hamiltonian diagonalized by the wavefunction defined in Eq. (10.76) is

$$H = \sum_\mu |\mu\rangle E_\mu \langle\mu| + \sum_{\mu\mu'} V_{\mu\mu'} = \sum_\alpha E_\alpha |\alpha\rangle\langle\alpha|. \qquad (10.77)$$

The question that arises is which interaction V should be used and how large the subspace should be to diagonalize it. A first answer to this question can be given by remembering that even at moderate excitation energies, the density of states $|\alpha\rangle$ is very high, of the order of 10^{20} levels per MeV for a nucleus of mass number equal to 200 at an excitation energy of 50 MeV. In a region of such a high density of levels it makes little sense to ask about the explicit properties of each individual state. Instead, the useful information relating to the nuclear structure is contained in the statistical properties of the many eigenstates.

Using the Gaussian Orthogonal Ensemble (GOE), one obtains a Gaussian distribution for the amplitude of the eigenvectors, that is,

$$P(X_\mu^\alpha) = \frac{1}{\sqrt{2\pi\sigma^2}} \exp\left(\frac{-X_\mu^\alpha}{2\sigma^2}\right), \tag{10.78}$$

leading to

$$\langle(X_\mu^\alpha)^2\rangle = \int dX_\mu^\alpha (X_\mu^\alpha)^2 = \sigma^2 = \frac{1}{N}, \tag{10.79}$$

where N is the dimension of the matrices. It is quite unlikely to obtain a fair estimate of N, that is, of the number of multi-quasiparticle states $|\mu\rangle$ that contribute to the compound state $|\alpha\rangle$. On the other hand, it is possible to estimate the range of energies Γ_μ over which the state $|\mu\rangle$ is distributed (cf. Eq. (9.13)). A simple parametrization of the corresponding strength is given by

$$\rho_\mu \langle(X_\mu^\alpha)^2\rangle = \frac{1}{2\pi} \frac{\Gamma_\mu}{(E - E_\mu)^2 + (1/2\Gamma_\mu)^2}, \tag{10.80}$$

where ρ_μ is the density of μ-states. The quantity $\hbar/\Gamma_\mu$ can be viewed as the hopping time needed for the compound nucleus to jump from one state $|(\alpha, \beta, I, U)\rangle$ to another, and consequently measures the time an excited nucleus can keep memory of a given deformation and orientation (cf. also Sect. 10.8.3).

The giant dipole resonance built upon the compound level defined in Eq. (10.76) is an identical set of levels (vibrational partner of the original states $|\alpha\rangle$) (cf. Figs. 6.9 and 10.3, also Fig. 1.3)

$$|\alpha(I), k\rangle = \Sigma_\mu X_\mu^\alpha |\mu(\beta, \gamma, \Omega; I); k\rangle, \tag{10.81}$$

displaced upwards by the giant dipole frequency. These states are the giant resonances implied by the Axel–Brink ansatz. In what follows, we identify with the label $0, \pm 1$, of the spherical components of the GDR with respect to the axis of nuclear rotation. The fact that the eigenstate $|\alpha(I)k\rangle$ contains the same amplitude X_μ^α as the compound state $|\alpha(I)\rangle$ ensures that the vibrational partner contains the full strength of the giant dipole resonance.

While the multi-quasiparticle states $|\mu\rangle$ form a complete set of states, the vibrational partners do not. They are lifted up from the states $|\mu\rangle$ into a region of excitation energy where they are embedded in a dense background of more complex states $|\lambda\rangle$, the density of which essentially corresponds to the total level density, implying that the set of states $\{|\mu, k\rangle, |\lambda\rangle\}$ is complete. The interaction of the GDR with the ensemble of states $|\lambda\rangle$ is governed by the coupling to doorway states containing an uncorrelated particle–hole excitation and a collective surface vibration on top of the excitations present in $|\mu\rangle$, as discussed in Sect. 4.2 (cf. Fig. 4.12).

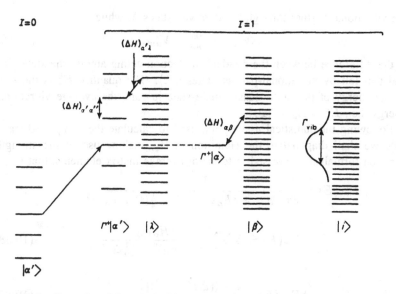

Figure 10.3. Illustration of the steps used in the calculation of the vibrational strength function. The compound nuclear states $|\alpha'\rangle$ (left) at spin $I = 0$ are excited to their vibrational partner $\Gamma^\dagger|\alpha'\rangle$ at spin $I = 1$. These states are embedded in a dense background of states $|\lambda\rangle$. The strength function of the state $\Gamma^\dagger|\alpha\rangle$ is obtained by first diagonalizing the Hamiltonian at spin I except for the couplings to the one state $\Gamma^\dagger|\alpha\rangle$, leading to the states $\beta\rangle$). The remaining couplings between $\Gamma^\dagger|\alpha\rangle$ and the set $|\beta\rangle$ lead to the strength function shown to the right, when $\Gamma^\dagger|\alpha\rangle$ is expanded on the stationary states $|i, I\rangle$. (From Broglia et al. (1988).)

In keeping with this discussion, we write the Hamiltonian associated with the vibrational partner states as the sum of two terms

$$H_1 = H_1^{(0)} + \Delta H, \qquad (10.82)$$

where

$$H_1^{(0)} = \sum_{\mu k} |\mu k\rangle E_\mu^k \langle \mu k| + \sum_{\mu \mu' k} |\mu' k\rangle V_{\mu \mu' k} \langle \mu k|, \qquad (10.83)$$

and

$$\Delta H = \sum_{\mu k} |\mu k\rangle \Delta \omega_k(\mu) \langle \mu k| \ + \ \sum_{\mu k \lambda} (|\lambda\rangle V_{\mu k, \lambda} \langle \mu k| + h.c.)$$

$$+ \ \sum_{\mu \mu' k} |\mu k\rangle \Delta V_{\mu \mu'}^k \langle \mu' k|. \qquad (10.84)$$

The quantity

$$\Delta \omega_k(\mu) = \hbar \omega_k(\mu) - \hbar \overline{\omega}_k, \qquad (10.85)$$

is the shift in the vibrational energy with respect to the energy of the GDR associated with the equilibrium deformation. The matrix elements $V_{\mu k, \lambda}$ couple the vibrational partner through the doorway states λ, while

$$\Delta V_{\mu\mu'}^k = V_{\mu\mu'}^k - V_{\mu\mu'}, \tag{10.86}$$

is the difference between the residual interaction acting among the states $|\mu\rangle$ and that felt by vibrational-partner states $|\mu k\rangle$. The quantity E_μ^k is the sum of the energy of the multi-quasiparticle state $|\mu\rangle$ and the average vibrational energy, i.e., $E_\mu^k = E_\mu + \hbar\bar\omega_k$.

Following the statistical assumption used to calculate the compound states $|\alpha\rangle$, we shall diagonalize the Hamiltonian H_1, making use of the strength function formalism expressed in terms of average matrix elements, that is,

$$P_{\alpha k}(E) = \frac{1}{2\pi} \frac{\Gamma_{\alpha k} + \Delta}{(\hbar\bar\omega_k + \delta E_{\alpha k} - E)^2 + \frac{1}{4}(\Gamma_{\alpha k} + \Delta)^2}, \tag{10.87}$$

$$\Gamma_{\alpha k}(E) = \Delta \sum_\beta \frac{\langle(\Delta H_{\alpha k, \beta})^2\rangle}{(E - E_\beta)^2 + (\frac{1}{2}\Delta)^2}, \tag{10.88}$$

$$\delta E_{\alpha k}(E) = \sum_\beta \frac{\langle(\Delta H_{\alpha k, \beta})^2\rangle(E - E_\beta)}{(E - E_\beta)^2 + (\frac{1}{2}\Delta)^2}. \tag{10.89}$$

The states $|\beta\rangle$ appearing in these expressions are obtained by diagonalizing the full Hamiltonian H_1 among all the states of the set $\{|\alpha'k'\rangle, |\lambda\rangle\}$, while excluding $|\alpha k\rangle$, as shown in Fig. 10.3.

The average squared matrix element appearing in the above equation can be written as

$$
\begin{aligned}
\langle(\Delta H_{\alpha k, \beta})^2\rangle &= \sum_\lambda \langle\Delta H_{\alpha k, \lambda}^2\rangle|\langle\lambda|\beta|\rangle|^2 \\
&+ \Sigma_{\alpha'k'}\langle(\Delta H_{\alpha k, \alpha'k'})^2\rangle|\langle\alpha'k'|\beta\rangle|^2.
\end{aligned} \tag{10.90}
$$

Because the density of states $|\lambda\rangle$ is essentially equal to the density of states ρ,

$$\sum_\lambda |\langle\lambda|\beta\rangle|^2 \approx 1. \tag{10.91}$$

Consequently, the first term contributing to $\langle(\Delta H_{\alpha k, \beta})^2\rangle$ can be written as

$$
\begin{aligned}
\langle(\Delta H_{\alpha k, \lambda})^2\rangle &= \sum_\mu \langle(X_\mu^\alpha)^2\rangle V_{\mu k, \lambda}^2 = \langle V_{\mu k, \lambda}\rangle_\alpha^2 \\
&\approx \frac{\Gamma_Q(k)}{2\pi\rho}, \quad (\rho_\lambda \sim \rho),
\end{aligned} \tag{10.92}
$$

where $\Gamma_Q(k)$ is the damping width associated with the coupling of the giant dipole resonance to the low-frequency quantal vibrations of the nuclear surface (cf. Eq. (10.62); cf. also Sect. 4.2).

In what follows, we shall calculate the second contribution to $\langle(\Delta H_{\alpha k, \beta})^2\rangle$. The squared matrix element can be written as

$$\langle(\Delta H_{\alpha k, \alpha'k'})^2\rangle = \delta(k, k')\{\langle\Delta\omega^2(k)\rangle_{\alpha\alpha'} + \langle(\Delta V)^2\rangle_{\alpha\alpha'}\}, \tag{10.93}$$

with

$$\langle \Delta\omega^2(k)\rangle_{\alpha,\alpha'} = \sum_\mu \langle (X_\mu^\alpha)^2\rangle \langle (X_{\mu'}^{\alpha'})^2\rangle \Delta\omega_k^2(\mu)\delta(\mu,\mu'). \tag{10.94}$$

and

$$\begin{aligned}
\langle (\Delta V)^2\rangle_{\alpha,\alpha'} &= \sum_{\mu\mu'} \langle (X_\mu^\alpha)^2\rangle \langle (X_\mu^{\alpha'})^2\rangle (\Delta V_{\mu\mu'})^2 \\
&\sim \frac{\Gamma_{\Delta V}(k)}{2\pi\rho_\alpha} \quad (\rho_\mu \sim \rho_\alpha \ll \rho).
\end{aligned} \tag{10.95}$$

The summation in the last expression can be explicitly calculated with the random matrix framework, leading to

$$\begin{aligned}
\langle\ &(\Delta H)^2_{\alpha k,\beta}\rangle \\
\approx\ &\frac{\Gamma_Q(k)}{2\pi\rho} + \frac{\Gamma_{\Delta V}(k)}{2\pi\rho}\int dE_{\alpha k}P_{\alpha k}(E_\beta + E_{\alpha k}) \\
+\ &\frac{\overline{\Delta\omega_k^2}}{2\pi\rho}\int dE_{\alpha' k}\frac{\Gamma_\mu}{(E_{\alpha k} - E_{\alpha' k})^2 + \Gamma_\mu^2}P_{\alpha' k}(E_\beta + E_{\alpha' k}).
\end{aligned} \tag{10.96}$$

In writing the above expression, the substitution of $\langle \Delta\omega^2(k)\rangle^2_{\alpha\alpha'}$ with the average value $\overline{\Delta\omega_k^2}$ independent of β and γ has been made. It should be noted is that $\int dE_{\alpha k}P(E_\beta + E_{\alpha k}) = 1$.

Substituting Eq. (10.96) in the expression for $\Gamma_{\alpha k}(E)$ one obtains

$$\Gamma_{\alpha k}(E) = \Gamma_Q(k) + \Gamma_{\Delta V} + \overline{\Delta\omega_k^2}\int dx\frac{E - x}{(E - x)^2 + \Gamma_\mu}P_{\alpha k}(x + \hbar\overline{\omega_k}). \tag{10.97}$$

where it has been assumed that the energies are measured from $E_\alpha = 0$. Similarly,

$$\delta E_{\alpha k}(E) = \overline{\Delta\omega_k^2}\int dx\frac{E - x}{(E - x)^2 + \Gamma_\mu}P_{\alpha k}(x + \hbar\overline{\omega_k}). \tag{10.98}$$

Although the model of the compound nucleus used above contains very little physical information except for the statistical (random matrix) hypothesis, it has proved instrumental to individualize the parameters controlling the damping process, namely the spreading width Γ_μ (hopping time $\approx \hbar/\Gamma_\mu$) of the compound state $|\alpha\rangle$ on the multi-quasiparticle states $|\mu(\beta,\gamma;I)\rangle$, and the dispersion $\delta\omega_k = \sqrt{\overline{\Delta\omega_k^2}}$ of the vibrational frequencies $\omega_k(\mu)$ of the GDR built on the ensemble of states $|\mu\rangle$. The nuclear structure enters into the evaluation of these parameters. One might be tempted to add to the list of parameters $\Gamma_Q(k)$ and $\Gamma_{\Delta V}(k)$. However, they are not particularly related to the finite temperature of the system, and appear also in the study of giant resonances based on the ground state. The quantity $\Gamma_{\Delta V}(k)$ is expected to be, as a rule, small.

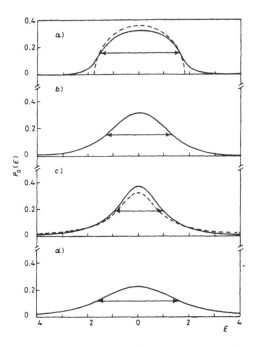

Figure 10.4. Examples of strength functions as obtained by solving Eqs. (10.87), (10.97), and (10.98) by iteration. a) $\Delta\omega_0 = 1$, $\Gamma_\mu = 0.2$, $\Gamma_Q = 0.1$; b) $\Delta\omega_0 = 1$, $\Gamma_\mu = 1$, $\Gamma_Q = 1$; c) $\Delta\omega_0 = 1$, $\Gamma_\mu = 2$, $\Gamma_Q = 1$; d) $\Delta\omega_0 = 1$, $\Gamma_\mu = 1$, $\Gamma_Q = 2$. The arrows point to half the maximum value, the distance between them yielding the FWHM. The thin dashed lines show the approximations to the functional form of the strength functions. In the top figure this is given by a semicircle, which is valid for Γ_μ, $\Gamma_Q \ll \Delta\omega_0$, and in the third figure a Breit–Wigner function, which is the analytic approximation when $\Gamma_\mu \gg \Delta\omega_0$. (From Lauritzen et al. (1988).)

The system of equations (10.87)–(10.89), or equivalently Eqs. (10.87), (10.97), and (10.98), can be solved by iteration. Fig. 10.4 shows examples of the resulting strength functions for different values of the parameters.

For small values of Γ_μ as compared to $\delta\omega_k$ and Γ_Q, both the statistical and quantal surface fluctuations can be treated as static deformations, and the GDR strength function approaches a Wigner's semicircle, displaying a FWHM equal to $\sqrt{12}\delta\omega_k$. In the other extreme, that is very fast hopping times ($\Gamma_\mu \gg \delta\omega_k$), the strength function approaches a Breit–Wigner shape of width equal to $\Gamma_Q + 2(\delta\omega_k)^2/\Gamma_\mu$ (cf. Fig. 10.4). In this second case, the GDR has a finite time to sample the ensemble of deformations associated with the states $|\mu\rangle$. Consequently, the spread in frequency $2\delta\omega_k$ is reduced by a fraction of time $\delta\omega_k/\Gamma_\mu$ available for the sampling. This is the reason why the solutions of Eqs. (10.87)–(10.89) display motional narrowing (cf. Kubo and Nagamiya (1969)), which is seen as a decrease in the FWHM as a function of Γ_μ, as well as a structural change from a confined strength-function (semicircle) to one with its wings stretched out (Breit–Wigner). Similar effects are expected in the case of the damping of rotational motion (cf. Ch. 11). A qualitative understanding of these results can be obtained by a simple inspection of Eqs. (10.87), (10.96), and (10.98):

1. $\Gamma_\mu(\Gamma_Q, \Gamma_{\Delta V}) \ll \delta\omega_k$. In this case,

$$
\begin{aligned}
\Gamma_{\alpha k}(E) &\sim 2(\delta\omega_k)^2 2\pi \int dx \delta(E - x) P(x + \hbar\overline{\omega}_k) \\
&= 4\pi(\delta\omega_k)^2 P(E + \hbar\overline{\omega}_\mu(k)) \\
&\sim 4\pi(\delta\omega_k)^2 P(E).
\end{aligned}
\tag{10.99}
$$

Substituting in Eq. (10.87) one obtains

$$
(E - \delta E_{\alpha k}(E))^2 + (\frac{1}{2}\Gamma_{\alpha k}(E))^2 = 2(\delta\omega_k)^2,
\tag{10.100}
$$

which is the equation of a circle. The strength function associated with this condition is Wigner's semicircle

$$
P_{\alpha k}(E) = \frac{2}{\pi^2 W} \sqrt{W^2 - E^2},
\tag{10.101}
$$

where $W = 2\delta\omega_k$. The corresponding FWHM is $\sqrt{12}\delta\omega_k$.

2. $\Gamma_\mu \gg \delta\omega_k$. In this case, Eq. (10.96) becomes

$$
\begin{aligned}
\Gamma_{\alpha k}(E) &\sim \Gamma_Q(k) + \Gamma_{\Delta V}(k) + \frac{2(\delta\omega_k)^2}{\Gamma_\mu} \int dx P(x + \hbar\omega_k) \\
&= \Gamma_Q(k) + \Gamma_{\Delta V}(k) + \frac{2(\delta\omega_k)^2}{\Gamma_\mu},
\end{aligned}
\tag{10.102}
$$

as mentioned above.

The rather high constancy of the GDR width observed in the Sn-isotopes at excitation energies ≥ 200 MeV (cf. Fig. 10.2), eventually testifies to the presence of motional narrowing in connection with the relaxation of the GDR in the compound nucleus (cf. also Sect. 10.8.3).

10.8 Analysis of the Experimental Data

In order to carry a detailed analysis of the strength functions and angular distributions associated with the γ-decay of the GDR from a compound nucleus, we shall discuss in what follows a simple, yet accurate model (Ormand et al. (1997a); cf. also Ormand et al. (1990) and (1990a), Alhassid and Bush (1990) and Alhassid (1994)).

Under the assumption of adiabatic motion, that is that the time scale associated with the thermal fluctuations is slow compared to that associated with the shift in the dipole frequency caused by the fluctuation, the GDR strength function is obtained as a weighted average over all shapes and orientations. Projecting the angular momentum J of the compound nucleus, the GDR cross-section is written as

$$
\sigma_{lab}(E) = Z_J^{-1} \int \frac{d\tau}{\mathfrak{I}^{3/2}(\beta, \gamma, \theta, \psi)} \sigma_{lab}(\vec{\alpha}, \omega_J; E)
$$
$$
\times \quad \exp[-F(T, \vec{\alpha}, \omega)/T] \qquad (10.103)
$$

where E is the photon energy, while

$$
Z_J = \int \frac{d\tau}{\mathfrak{I}^{3/2}} \exp[-F(T, \vec{\alpha}, \omega)/T].
$$

The fixed rotational frequency is determined by the average angular momentum of the system. The free energy F was discussed in Sect. 8.2. The cross-section $\sigma_{lab}(\vec{\alpha}, \omega_J; E)$ appearing in the integrand, is the projection in the laboratory frame of the corresponding quantity calculated in the intrinsic frame. This is obtained by diagonalizing a dipole Hamiltonian supplemented by the Coriolis and centrifugal forces, that is,

$$
H_D = \sum_{k=1,3} (p_k^2 + E_k^2 d_k^2) - \vec{\alpha} \cdot (\vec{d} \times \vec{p}) \qquad (10.104)
$$

where d_k is the giant-dipole operator associated with the dipole vibration along the axis k and p_k is the conjugate momentum. The resonance energies E_k are as in Eq. (10.60). Having obtained the eigenvalues and eigenvectors, labelled by ν, the cross-section in the intrinsic frame is written as

$$
\sigma_{int}(\vec{\alpha}, \omega; E) = \frac{4\pi^2 e^2 \hbar}{3mc} \frac{2NZ}{A} \sum_{\mu, \nu} |\langle \nu | d_\mu | 0 \rangle|^2 E
$$

$$
\times [BW(E, E_\nu(\alpha, \omega), \Gamma_\nu) - BW(E, E_\nu(\alpha, \omega), \Gamma_\nu)], \qquad (10.105)
$$

where $BW(E, E_a, \Gamma)$ is the Breit–Wigner function

$$
BW(E, E_a, \Gamma) = \frac{1}{2\pi} \frac{\Gamma}{(E - E_a)^2 + \Gamma^2/4}. \qquad (10.106)
$$

The widths Γ_ν are obtained from the spherical value Γ_0 according to the empirical relation (Carlos et al. (1974))

$$
\Gamma_\nu = \Gamma_0 \left(\frac{E_\nu}{E_0}\right)^\delta \qquad (10.107)
$$

with $\delta \approx 1.8$ and $E_0 = E_D \approx 80A^{-1/3}$ MeV, (cf. Eq. (2.1)) and $\Gamma_0 \equiv \Gamma_{GDR}^{\downarrow}$ (cf. Sects. 9.2 and 9.3, and Fig. 9.7; cf. also Fig. 6.6).

It should be noted is that for excitation energies of few hundreds of MeV, the width $\Gamma_{CN}^{\uparrow}$ for the decay of a compound nucleus by particle emission (cf. Eq. (5.7)) is of the order of an MeV, and should be added to Γ_ν twice, corresponding to the particle decay width of the initial and final states coupled by the dipole gamma-decay (Chomaz (1995a)). This is a general prescription of quantum theory to describe the decay connecting unstable states (Berestetskii et al. (1976)).

10.8.1 The nuclei ^{120}Sn and ^{208}Pb: the role of temperature

Inelastic scattering of α-particles at bombarding energy of 40 and 50 MeV/u has been measured in coincidence with γ-decay. The amount of angular momentum transferred to the nucleus is small (less than 20 $\hbar$ at the highest excitation energies) as compared to that associated with fusion reaction induced by heavy ions. These data thus provide a test of the ability theory has to correctly predict the variation of the FWHM of the GDR as a function of temperature, essentially without any contribution arising from angular momentum. The experimental findings (Ramakrishnan et al. (1996), that is the FWHM of the GDR strength function for both ^{120}Sn and ^{208}Pb as a function of temperature, are shown in Fig. 10.5. The results of theoretical calculations are also displayed in these figures. The calculations were carried out as discussed at the beginning of this section, making use of the free energy shown in Fig. 8.7, and setting $E_0 = 15$ MeV and $\Gamma_0 = 5$ MeV for ^{120}Sn and $E_0 = 13.7$ MeV and $\Gamma_0 = 4$ MeV for ^{208}Pb. The shell correction in ^{120}Sn was found to be quite small and effectively can be ignored.

The solid lines in Fig. 10.5 correspond to theoretical values obtained with zero angular momentum. The dependence of the FWHM for both nuclei on angular momentum is illustrated in Fig. 10.6, where it is seen that for $J \leq 25\hbar$, the FWHM is essentially unchanged from the $J = 0$ value. Given that the largest average angular momentum in the system studied is of the order of 20 $\hbar$, the effects due to angular momentum are expected to be negligible.

The dependence of the FWHM on temperature is quite different between ^{120}Sn and ^{208}Pb. The FWHM in ^{208}Pb appears to be suppressed at lower temperatures. This is accounted for in the theoretical model by the rather strong shell corrections in ^{208}Pb that favor the spherical shape at low temperatures. The effect of such strong shell corrections is to limit the influence of thermal fluctuations at low temperatures, thereby reducing the observed width. This is also illustrated in Fig. 10.6, where the dotted line in the panel for ^{208}Pb indicates the FWHM obtained assuming no shell corrections. Furthermore, the quantity $2\Gamma_{CN}^{\uparrow}$ was added to the calculated FWHM, to take into account the finite escaping width displayed by the initial and final compound nucleus states connected by the GDR transition (cf. Fig. 10.5).

10.8.2 The nucleus ^{110}Sn: the role of angular momentum

The investigation of the angular momentum dependence of the GDR width at a fixed temperature provides a test for the ability theory has to account for the variations of the FWHM as a function of rotational frequency. Results of such a test carried out (Mattiuzzi et al. (1997); cf. also Ormand et al. (1990)) for nuclei belonging to mass regions A $\approx$ 110 and A$\approx$ 170 are shown in Figures 10.7 and 10.8.

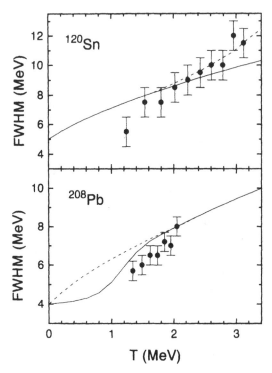

Figure 10.5. The FWHM of the GDR strength function as a function of temperature for ^{120}Sn and ^{208}Pb. Experimental data are represented by filled circles, while the solid lines represent the theoretical values obtained at zero angular momentum. For ^{208}Pb the dotted line is the FWHM obtained assuming no shell corrections. For ^{120}Sn and ^{208}Pb, the dashed line represents the FWHM obtained by including the increase to the intrinsic width due to the evaporation of particles from the compound system. (From Ormand et al. (1997a).)

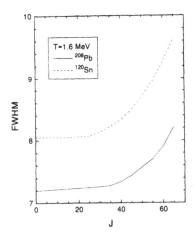

Figure 10.6. The FWHM in ^{120}Sn (dashed line) and ^{208}Pb (solid line) at $T = 1.6$ MeV as a function of angular momentum. (From Ormand et al. (1997a).)

In the case of Sn isotopes the width of the GDR is found to be roughly constant for angular momenta smaller than $35\hbar$ and to increase rapidly at higher values of the angular momenta. As the rotational frequency becomes larger, the nucleus undergoes an oblate flattening due to centrifugal effects. This is characterized by a shift in the minimum of the nuclear free energy to oblate non-collective shapes ($\gamma = 60°$, cf. Fig. 6.4), whose magnitude is determined by the interplay between the surface and rotational energies, and is sensitive to the moment of inertia of the system. Because of the dependence in the free energy on the inverse of the moment of inertia, one finds that for nuclei such as ^{106}Sn, the equilibrium deformation, β_{eq}, increases rapidly with angular momentum and, as a consequence, the total GDR strength function undergoes a further splitting, which increases the FWHM. It should also be pointed out, however, that even though the equilibrium deformation increases with angular momentum, an increase in the FWHM of the GDR does not occur until the equilibrium deformation increases sufficiently so as to affect the thermal average value of $\langle \beta \rangle$ $(= Z^{-1} \int d\tau \beta \exp(-F/T))$. In addition to the mean value, the nucleus also experiences a spread in deformations, which can be measured by the variance $\Delta\beta = \sqrt{\langle \beta^2 \rangle - \langle \beta \rangle^2}$. As long as $\beta_{eq} < \Delta\beta$, neither the mean deformation $\langle \beta \rangle$ nor the FWHM of the resonance increase significantly. For the most part, at low angular momentum, the effect of the equilibrium deformation is "washed out" by the thermal fluctuations. This is illustrated in the top panel of Fig. 10.7, where, for ^{106}Sn, β_{eq} is compared with the variance $\Delta\beta$. The increase in the mean value of β as a function of angular momentum is shown in the middle panel of Fig. 10.8.

In the bottom panel of Fig. 10.8, the FWHM obtained from theoretical calculations for ^{106}Sn at $T = 1.8$ MeV are compared with experimental data (solid line). For the most part, theory and data are in overall agreement. In particular, the constancy of the GDR–FWHM up to angular momentum $J \sim 35\hbar$ is well reproduced. Indeed, by comparing the middle and bottom panels, a clear correlation between the FWHM and $\langle \beta \rangle$ is seen. At higher angular momenta, the theoretical values appear to be systematically smaller than the experimental values. It should be pointed out however, that the results of the data are somewhat sensitive to the choice of the level-density parameter a, and as such the extracted temperatures are uncertain by a few hundred keV. In addition, the experimental data were obtained by fitting to a single Lorentzian, while the theoretical cross-sections comprise a sum of several Lorentzians. A further supposition may lie in the moment of inertia being somewhat smaller than the rigid-body values. As such, we repeated the calculation while setting the radius parameter $r_0 = 1.1$ fm (dotted line), which represents a 16% reduction in the moment of inertia. We find that the FWHM is increased slightly at higher values of angular momentum, in better agreement with experiment.

Following the arguments outlined above regarding the angular momentum dependence of the FWHM, it should then be expected that for heavier nuclei, such as ^{176}W or ^{208}Pb, the effect of oblate flattening should be reduced. As a consequence the FWHM should exhibit less dependence on angular momentum. This is clearly exhibited in the top panel of Fig. 10.8, where the experimental FWHM for ^{176}W at $T \approx 1.5$ MeV is seen to be essentially constant as a function of angular momentum. This is also in overall agreement with theoretical predictions, as is illustrated by the dashed line in the figure.

To further illustrate the influence of the moment of inertia on the FWHM of the GDR as a function of angular momentum, we compare (in the bottom panel of Fig. 10.8) the FWHM for the nuclei ^{106}Sn, ^{120}Sn, ^{176}W, and ^{208}Pb as a function of angular momentum at temperatures of the order 1.5–1.8 MeV. In the figure, it is seen that the conspicuous increase in the FWHM on angular momentum exhibited by the GDR of ^{106}Sn as a function of angular momentum is significantly abated in the case of the GDR in heavier nuclei. In particular, in the case of in ^{120}Sn, which, with its 14 additional neutrons, has a moment of inertia that is approximately 15% larger than ^{106}Sn, the increase in the FWHM at $60\hbar$ is nearly 2 MeV smaller. Again, the correlation with the increase in $\langle \beta \rangle$ shown in the middle panel of Fig. 10.8 for each of the nuclei under discussion is evident.

Before concluding this section it is important to recall that the model describes well also the values of the $a_2(E_\gamma)$ obtained from the angular distribution measurements. Contrary to the strength function data that is obtained through a statistical model analysis of the data, the $a_2(E_\gamma)$ data can be directly compared to the model predictions. The results for the $a_2(E_\gamma)$ are shown together with calculations in Sect. 6.1.3.

10.8.3 The nucleus ^{92}Mo: motional narrowing

An analysis of the observed strength function and of the angular distribution in ^{92}Mo (Gundlach et al. (1990)) have been carried out allowing for fluctuations in both the nuclear shape and the nuclear orientation and allowing for the system to spend a finite amount of time at each configuration in the (β, γ, Ω)-space (Ormand et al. (1992)). The time dependence of the fluctuations were taken into account using a model based on the Kubo–Anderson process (cf. Dattagupta (1987) and references therein), In this model, jumps are made between the various deformations and orientations independent of the initial starting point, and the conditional probability $P(\vec{\alpha}, t, \vec{\alpha}_0, t_0)$ of having deformation $\vec{\alpha} \equiv (\beta, \gamma, \Omega)$ at time t, after having been at the point $\vec{\alpha}_0$ at t_0, is

$$\begin{aligned} P(\vec{\alpha}, t, \vec{\alpha}_0, t_0) &= \exp[-\lambda(t - t_0)]\delta(\vec{\alpha} - \vec{\alpha}_0) \\ &+ (1 - \exp[-\lambda(t - t_0)])p(\vec{\alpha}). \end{aligned} \quad (10.108)$$

where λ is the mean jumping rate and $p(\vec{\alpha})$ is the stationary probability distribution $Z^{-1} \exp[-F/T]$. As λ increases, the effects of the thermal fluctuations on the line shape, as well as on the angular distribution are mitigated in keeping with the phenomenon of motional narrowing discussed in Sect. 10.7.

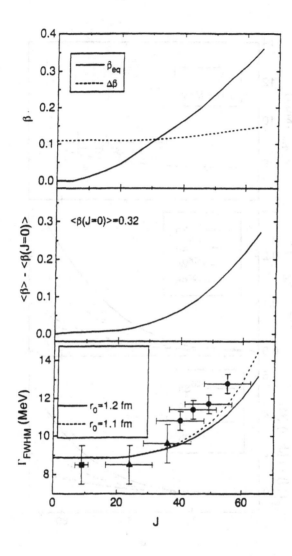

Figure 10.7. Top panel: Evolution of the equilibrium deformation (full drawn line) for ^{106}Sn at $T = 1.8$ MeV as a function of the angular momentum $\langle J \rangle$. The dashed line shows the value of $\Delta\beta$ defined in the text. Middle panel: The mean nuclear deformation with respect to that at zero angular momentum is shown. Bottom panel: The measured values of the GDR width from Mattiuzzi et al. (1997) (filled triangles) and those from Ramakrishnan et al. (1996) (filled square) and from Bracco et al. (1995) (filled circles) are compared with predictions of the thermal fluctuation model. The horizontal bars represent the full-width half-maximum of the spin distribution associated with the experimental data. Two assumptions were made for the moment of inertia of the nucleus, one that is equal to that of a rigid rotor (full drawn line), and the other that is 16% smaller (dashed line). (From Mattiuzzi et al. (1997).)

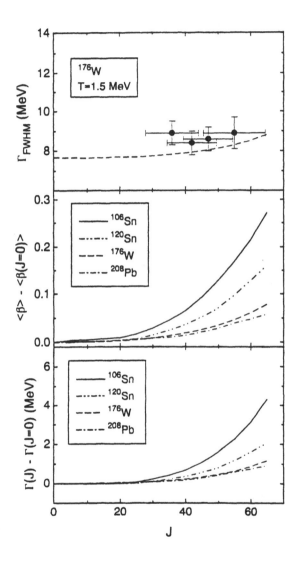

Figure 10.8. Top panel: Evolution of width of the giant dipole resonance in ^{176}W at $T = 1.5$ MeV. The data points are compared to the predictions of the thermal fluctuation model described in the text. The horizontal bars represent the full-width half-maximum of the spin distribution associated with the experimental data. Middle panel: The mean nuclear deformation with respect to that at zero angular momentum for different nuclei (for Sn isotopes $T = 1.8$ MeV while for ^{176}W and ^{208}Pb $T = 1.5$ MeV). Bottom panel: Evolution of the GDR width with angular momentum as predicted by the thermal fluctuation model. The plotted values are relative to zero angular momentum and are given for different nuclei. (From Mattiuzzi et al. (1997).)

Coupling the relation (10.108) to Eqs. (10.104)–(10.107), a χ^2-fit on the ^{92}Mo data (strength function and angular distribution) was performed, allowing the intrinsic width Γ_0, the total oscillator strength S_0 carried by the resonance normalized to the dipole EWSR, and the jumping rate $\lambda_{\beta,\gamma}$ associated with shape (β, γ)-fluctuations and λ_Ω associated with orientation fluctuations, to vary. The minimum was found for $\Gamma_0 = 6.8$ MeV (close to the width of the GDR based on the ground state), $S_0 = 1$, $\lambda_{\beta,\gamma} = 7.5$ MeV and $\lambda_\Omega = 0.75$ MeV. The results thus seem to indicate both the presence of motional narrowing, as well as of different time scales controlling fluctuations in shape (fast) and in orientation (slow).

11

Rotational Motion

It is well established that closed shell nuclei are spherical in their ground state. Adding nucleons polarizes the closed shells and eventually the polarization effects may become so large that the mean field becomes deformed. In other words, the mean field solution of the nuclear many-body system displaying the lowest energies may not be spherically symmetric. As a rule, these solutions display quadrupole deformation. This is the phenomenon of spontaneous symmetry breaking of rotational invariance that makes it possible to specify an orientation of the system in ordinary space. Restoration of symmetry arises from the rotation of the system as a whole, as evidenced by the presence of rotational bands. In fact, rotations represent the collective mode associated with such a spontaneous symmetry breaking phenomenon.

11.1 Cranking Model

The energy of members of an axially symmetric nucleus rotating with spin I is

$$E(I) = \frac{\hbar}{2\mathcal{J}}I(I+1), \tag{11.1}$$

where $\mathcal{J}$ is the moment of inertia. This quantity can be calculated summing up the inertial effect of each particle as it is dragged by a uniform rotating potential (cf. Fig. 11.1). In this approach the potential appears to be externally "cranked." The cranking Hamiltonian

$$H = H_0 - \hbar\omega J_x, \tag{11.2}$$

describing the particle motion in a potential rotating with frequency ω about the x-axis is obtained from the single-particle Hamiltonian H_0 (Nilsson Hamiltonian, Nilsson (1955)), describing the motion of a nucleon in the fixed deformed mean field, by the addition of the term proportional to the x-component of the total angular momentum. This term represents the Coriolis force acting on

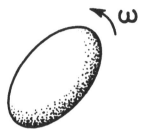

Figure 11.1. Schematic representation of the rotation of the single-particle potential with frequency ω.

the rotating coordinate frame fixed to the nucleus. The moment of inertia is obtained from a second-order perturbation treatment of this term

$$\mathcal{J} = 2\hbar^2 \sum_i \frac{|\langle i|J_x|0\rangle|^2}{E_i - E_0}, \tag{11.3}$$

and involves a sum over the excited states i. For independent particle motion, this expression leads to a value approximately corresponding to rigid rotation. The observed value of $\mathcal{J}$ for small values of the rotational frequency are, on the other hand, a factor of 2–3 smaller than the rigid moment of inertia (cf., e.g., Bohr and Mottelson (1975) and refs. therein). This is because nucleons in a nucleus interact through pairing forces. This interaction scatters pairs of particles moving in time-reversal states and carrying angular momentum zero (cf. Fig. 11.2), leading to a condensate of Cooper pairs. This phenomenon is closely related to that found at low temperatures in superconductors, a phenomenon which is accurately described by the BCS equations (Bardeen et al. (1957), Schrieffer (1964), Bohr et al. (1967)).

Pairing correlations affect the ability of nucleons to generate angular momentum. In fact, insofar as the pairs are coupled to spin zero, they can contribute nothing to the angular momentum of the rotating system. This causes a conspicuous reduction of the nuclear moment of inertia, as required by the experimental findings. It follows that angular momentum will tend to weaken the pairing correlations, thus increasing the moment of inertia and reducing the rotational energy.

The mechanism of this weakening is the Coriolis force, which acts oppositely on the two members of the Cooper pairs, lifting the degeneracy. In fact, the term $\hbar\omega J_x$ in Eq. (11.2) violates time reversal invariance. Ultimately the Coriolis force wants to align the particle angular momentum as well as possible with the rotational axis, as illustrated in the bottom of Fig. 11.2. This process is analogous to the effect a magnetic field has on the paired electrons of a superconductor. However, the nuclear phase transition is not sharp as in the case of an infinite system but gradual, as evidenced by a smooth rise of the moment of inertia for spins up to 20–30 $\hbar$ (cf., e.g., Shimizu et al. (1989), Broglia (1988), Lauritzen et al. (1993) and refs. therein). But, within

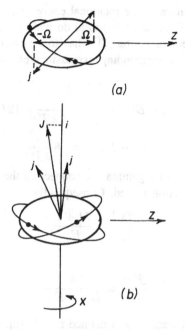

Figure 11.2. The coupling schemes in deformed nuclei. In the absence of rotation (top) particles with angular momentum j are in time reversed orbits with projection $\pm\Omega$ along the symmetry (z) axis. At high rotational frequencies the particles couple to angular momentum J, aligned as well as possible with the rotational (x) axis, along which they have projection i. (After Stephens (1985).)

this gradual rise there is occasionally a large irregularity that corresponds to the rather complete alignment of a particular pair of high-j nucleons. This comes about because the Coriolis force is proportional to j and thus affects high-j particles most strongly, as for example those moving in a $i_{13/2}$ orbital. Just as a toy gyroscope will attempt to align its rotation axis with that of its rotating frame, so a pair of high-j particles tends to align its rotation axis (angular momentum) with that of the rotating nucleus, thereby decreasing its energy relative to a rotational band without such alignment. This behaviour is now well studied for many nuclei and the change described above corresponds to a crossing of two rotational bands (Stephens and Simon (1972)). A band with two aligned $i_{13/2}$ neutrons crosses the ground-state band, which has all the nucleons around the Fermi energy participating in the pairing correlation (BCS ground state). Thus a discontinuity in the moment of inertia actually corresponds to a shift into another band.

As a consequence of the interplay between collective and single-particle motion, there are a variety of moments of inertia one can extract from the experimental data. The basic distinction to be made is between kinematic and dynamic values (Bohr and Mottelson (1981), Shimizu et al. (1990)). These two moments of inertia are labelled $\mathcal{J}^{(1)}$ and $\mathcal{J}^{(2)}$ and are connected to the first and second derivative of the rotational energy with respect to spin. The "kinematic" moment of inertia $\mathcal{J}^{(1)}$ has to do with the motion of the system. The "dynamic" moment of inertia $\mathcal{J}^{(2)}$ has to do with the way the system reacts to a force. The corresponding expressions can be obtained from Eq. (11.1) leading to

$$E_\gamma = E(I) - E(I - 2) \quad = \quad \frac{\hbar^2}{2\mathcal{J}^{(1)}} 2(2I - 1),$$

$$E_{\gamma_1} - E_{\gamma_2} \quad = \quad 8\frac{\hbar^2}{2\mathcal{J}^{(2)}}, \tag{11.4}$$

where E_γ is the energy of a gamma-ray connecting the members with spin I and spin $I - 2$ of a rotational band. Consequently

$$\frac{\mathcal{J}^{(1)}}{\hbar^2} = \frac{2I + 1}{E_\gamma}, \tag{11.5}$$

and

$$\frac{\mathcal{J}^{(2)}}{\hbar^2} = \frac{4}{E_{\gamma_1} - E_{\gamma_2}}. \tag{11.6}$$

Because the rotational frequency is defined as $\frac{E_\gamma}{2\hbar}$, Eq. (11.5) corresponds to the classical relation $\mathcal{J}^{(1)}\omega = I\hbar$. On the other hand, the relation

$$\mathcal{J}^{(2)}\omega = R = I - i, \tag{11.7}$$

expresses the fact that the total angular momentum I is shared between a collective rotational part R, and the nucleon alignment component i.

11.2 Warm Nuclei

In heavy-ion reactions leading to fusion, large amounts of angular momentum and temperature can be imparted to the compound nucleus. In analogy to the cooling process of a liquid drop, nucleon evaporation can reduce the temperature of a nucleus, without much changing its angular momentum content. In fact, the energy loss per evaporated nucleon varies as the corresponding separation energy, while angular-momentum loss per evaporated nucleon varies as I/A, where I is the total angular momentum and A the number of nucleons in the rotating system. Thus the excitation energy can be reduced to a point where further evaporation is no longer possible, but the resulting "evaporation residue" still retains a very large amount of angular momentum. There will be an appreciable excitation energy in the residue, although the system is not very hot, in the sense that a large fraction of this energy is tied up in the single degree of freedom associated with the angular momentum. Typical values of the angular momentum and of the thermal energy, that is the excitation energy above the yrast line, are 50–60 units of $\hbar$ and 2–3 MeV respectively, thus implying temperatures of the order of 0.5 MeV.

In keeping with the discussion above, and making use of the relations given in Eqs. (11.1) and (11.6), one expects that the gamma-decay of warm, rotating nuclei will lead to a regular pattern of valleys and ridges along the two dimensional $(E_{\gamma_1} - E_{\gamma_2})$-coincidence spectrum, as shown in the schematic diagrams (a), (b) and (c) of Fig. 1.12. The pattern will hardly be altered taking into account the phenomena of band crossing and of fluctuations in the moment of inertia discussed above (cf. Figs. 1.12(d), (e) and (f)).

Experimentally the situation is quite different (cf. Fig. 1.13). Although one observes a pattern of valleys and ridges, the correlation in the energy of two successive gamma-transitions is much weaker than that associated with a perfect rotor (Herskind (1984), Draper et al. (1986)). This is because the dominant part of the enhanced electric quadrupole decay passes through regions of high level density. The residual interaction mixes these simple basis states, leading to complex compound nuclear states, eigenstates of the nuclear Hamiltonian. Each state is then spread over many compound nucleus eigenstates. The energy interval over which this phenomenon takes place is known as the damping width $\Gamma_{CN}^{\downarrow} \equiv \Gamma_\mu$, of the intrinsic states (cf. Eq. (9.13)). If all the intrinsic states behaved equally as a function of the rotational frequency, the quadrupole rotational decay pattern would correspond to discrete, sharp transitions, unaffected by the residual interaction. However, because each intrinsic configuration responds differently to the effect of Coriolis and centrifugal forces, the rotational bands at a given spin will display a dispersion of frequencies $\Delta\omega$. Consequently, rotational bands which at spin I are situated in a narrow energy interval, are spread out in energy at spin $I - 2$, implying damping of the rotational motion. The residual interaction at both spins I and $I - 2$ will lead to a pattern of quadrupole rotational transitions which, starting at spin I, populates a distribution of states at spin $I - 2$. The width Γ_{rot} of the associated strength function is the rotational damping width (Lauritzen et al. (1986), Leander (1982), Liotta and Sorensen (1978)). Although the situation seems to be somewhat more subtle (cf. Sect. 11.5.2), these transitions can be correlated with the intensities observed in the valleys, while the ridges are built mainly out of discrete electric quadrupole transitions between states belonging to rotational bands close to the yrast band.

Simulation calculations (Bracco et al. (1996)) of the rotational decay carried out are compared with the experimental data in Fig. 11.3. From this comparison, values of Γ_{rot} and U_0, that is, the energy above yrast (heat energy) where rotational damping starts taking place can be inferred. Values of the order of 200 keV and 1 MeV, respectively, are not inconsistent with the experimental findings.

11.3 Rotational Damping and NMR: An Analogy

The nuclear rotational damping in the compound nucleus has an analog in the linewidth observed in the phenomenon of nuclear magnetic resonance (NMR)

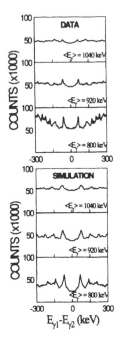

Figure 11.3. Experimental and simulated spectra are compared for nucleus ^{168}Yb. Each spectrum is obtained from a γ–γ coincidence matrix projecting along lines perpendicular to the main diagonal. The average transition energies are $(E_{\gamma 1} + E_{\gamma 2})/2 = 800$ keV for the bottom part, 920 keV for the middle part and 1040 keV for the top part. In all cases the cuts are of width of 60 keV. The data are corrected for Compton background and for the energy dependent detector efficiency. The simulated spectra were obtained using at each spin value a set of 1000 levels calculated with the cranked shell model plus residual interaction. In the simulation of the decay path of the nucleus the level density and the B(E2) strength were deduced from the calculated levels, and the B(E1) strength is fixed to reproduce the average experimental flow of the decay. The energy resolution of the experimental data has been used also for the simulation. (From Bracco et al. (1996).)

in condensed matter (cf., e.g., Slichter (1963)). Each nuclear spin in its lattice site precessing with a Larmor frequency determined by the external magnetic field B_0 is the analog of a rotational band. The whole crystal at a given temperature, and with the magnetic-dipole interaction between nearest neighbours taken into account, is the analog of the mixed-band eigenstates of the compound nucleus. The spin system is probed by an external time-dependent magnetic field perpendicular to the constant field B_0, while the analogous information on the dynamics of the nuclear rotational motion is contained in the emitted electric-quadrupole radiation. The linewidth observed in the absorption process is due to the spread in Larmor frequencies and reflects the spatial inhomogeneity of the local magnetic field. Its analog in the nuclear case is the rotational-damping width Γ_{rot} of the electromagnetic quadrupole decay, which has its origin in the spread in the rotational frequencies of the manifold of bands (Broglia et al. (1987)).

While the damping width of collective motion of a many-body system is, as a rule, an increasing function of the temperature, the nuclear magnetic resonance-line becomes narrower with increasing T. This phenomenon, known as motional narrowing, arises because higher temperatures imply shorter diffusion times between different lattice sites. Time inhomogeneity is able to average out spatial inhomogeneity. It is noted that motional narrowing occurs quite generally in systems where the frequency of a resonant process may be influenced by time-dependent perturbations (cf. Sects. 10.7 and 10.8.3).

From the above discussion, it emerges that the rotational-damping width Γ_{rot} is controlled by two quantities: the spread in rotational frequencies $\Delta\omega_0$ of the unperturbed states and the time (hopping time) a compound-nucleus state stays in any single rotational band (diffusion time for atoms to move between lattice sites). This last quantity is connected with the range of energies over which the rotational bands are mixed by the residual interaction ($\tau = \frac{\hbar}{\Gamma_\mu}$), cf. Eq. (9.13)).

In keeping with the analogy to NMR, the rotational-damping width Γ_{rot} can be written as $\Gamma_{rot} = \frac{2\hbar}{T_2}$, where T_2 is the time it takes for the different bands (the nuclear spins to NMR) to lose phase coherence (dephasing time). If the spread in frequencies $\Delta\omega_0$ is large compared with the diffusion rate $(2\Delta\omega_0 > \tau^{-1})$, dephasing results directly from the differences in rotational frequencies:

$$T_2 = (2\Delta\omega_0)^{-1}, \quad \text{if } 2\Delta\omega_0 > \tau^{-1}. \tag{11.8}$$

The factor of 2 results from the quadrupole nature of the radiation. In the opposite extreme of very short diffusion time, the dephasing results from a random walk over $\frac{T_2}{\tau}$ different sites and thus

$$T_2 = (2\Delta\omega_0)^{-2}\tau^{-1}, \quad \text{if } 2\Delta\omega_0 \gg \tau^{-1}. \tag{11.9}$$

The above relations applied to the nuclear rotational damping imply

$$\Gamma_{rot} = \begin{cases} 2(2\hbar\Delta\omega_0), & \text{if } \Gamma_\mu < 2\Delta\omega_0, \\ 2(2\hbar\Delta\omega_0)^2/\Gamma_\mu, & \text{if } \Gamma_\mu > 2\Delta\omega_0. \end{cases} \tag{11.10}$$

The spreading width Γ_{rot} of the individual configurations in the compound nucleus is the analog of the diffusion time in NMR and will, in general, depend on the heat energy U of the states considered.

Before concluding this section it is important to emphasize that the rotational damping width Γ_{rot} and the compound width Γ_μ have quite different origins and are thus, unrelated to each other. The quantity $\hbar/\Gamma_\mu$ measures the time the compound nucleus spends in a simple shell model configuration, while the quantity $\hbar/\Gamma_{rot}$ measures the time it takes for a compound nucleus eigenstate at spin I, to get distributed over a range of compound nucleus eigenstates at spin $I-2$, still collecting the full rotational transition strength. In other words, at finite temperature any particular intrinsic eigenstate can be represented as a random superposition of many different configurations. The individual configurations in this combination respond differently to the Coriolis force resulting from the nuclear rotation, depending essentially on the difference in value of the aligned angular momentum of each configuration (cf. Eq. (11.7)). This implies that on changing the rotational frequency slightly, as the system moves from spin I to spin $I - 2$, the different configurations in the given compound nucleus state will become somewhat out of phase with each other and will no longer represent an eigenstate of the intrinsic Hamiltonian. Rather, the particular combination that was an eigenstate at I will represent a wave packet of eigenstates at $I - 2$, and it is the energy width of this wave packet that is observed as the rotational damping width. Rotational damping reflects the varying effect of the Coriolis force on the compound states as a function of rotational frequency. On the other hand, the spreading width Γ_μ of the compound eigenstates measures the energy interval over which a "typical" shell model configuration couples to compound nucleus eigenstates. Consequently, it is a local quantity depending only on the properties of the system at fixed values of the parameters determining the physical state of the nucleus (e.g., spin and excitation energy).

11.3.1 Ordered and chaotic motion

Rotational damping, as well as the mean energy of rotational transitions, depends on the details of the Hamiltonian. On the other hand, compound nucleus damping is expected to be insensitive to the particular details of the Hamiltonian. It will lead, once diagonalized, to universal fluctuations of intensities and energies. In principle, this part of the Hamiltonian can be represented by random matrix Hamiltonian, in particular that associated with the Gaussian Orthogonal Ensemble (GOE). Quite independently from Γ_{rot}, the quantity Γ_μ controls, together with the average level spacing $D = \varrho^{-1}$, ϱ being the density of levels, the range of validity of concepts like chaos and the associated presence of GOE level statistics and Porter–Thomas intensity fluctuations. It is expected that most of the experimental information on the damping of rotational motion arises from regions of the spectrum where (cf. Figs. 1.13 and 11.4):

a) $\Gamma_{rot} < \varrho^{-1}$ and $\Gamma_\mu < \varrho^{-1}$, that is, in the case of discrete rotational transitions giving rise to ridge-like structures in the γ–γ coincidence spectra.

b) $\Gamma_{rot} > \Gamma_\mu > \varrho^{-1}$ ($\Gamma_\mu > \Gamma_{rot} > \varrho^{-1}$), that is, rotational transitions displaying damping without (with) motional narrowing, and leading to valley-like regions in the γ–γ coincidence spectra.

The energies of the rotational transitions associated with regime a) will depend on the particular intrinsic configurations of the associated rotational bands. In such a situation, the level statistics (nearest neighbour spacings and Δ_3-statistics, also known as "spectral rigidity"), are expected to follow a Poisson distribution typical of ordered motion. On the other hand, in the regime b) of strongly interacting bands, the expectation for the level statistics and intensities is that associated with the GOE. That is, Wigner-like behaviour for

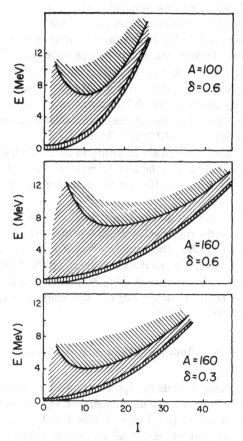

Figure 11.4. Schematic division of the rotational bands: (upper) discrete bands ($\Gamma_{rot} < \rho^{-1}$, $\Gamma_\mu < \rho^{-1}$, (middle) damped bands $\Gamma_{rot} > \Gamma_{rot} > \rho^{-1}$); (lower) compound bands $\Gamma_{rot} < \rho^{-1} < \Gamma_\mu$). (From Broglia et al. (1998).)

level spacings and Δ_3-statistics, and Porter–Thomas fluctuations for rotational intensities, typical of chaotic motion.

In this connection one can remember that the eventual successful analysis of the high resolution electron scattering data for the excitation of the isoscalar giant quadrupole resonance in ^{208}Pb (Kilgus et al. (1987)) discussed in Sect. 4.2.1, was based on assumptions like those discussed in relation with the regime b). This in spite of the fact that the width of the giant resonance is controlled by the coupling to specific doorway states (2p–2h configurations containing a surface vibration). However, at an excitation energy of the order of 10–11 MeV, the intensity distribution of the associated level spectrum has to display a Porter–Thomas distribution even if the damping width of the resonance is essentially controlled by the coupling to doorway states (cf. Sect. 4.2.2 and Fig. 4.18).

It should be noted, however, that situations where c) $\Gamma_{rot} < \varrho^{-1} < \Gamma_\mu$ could also be found along the rotational decay path (Mottelson (1993), (1993a), Åberg (1996), Broglia et al. (1987)). In this case the intrinsic states will exhibit Porter–Thomas intensity fluctuations, except for the collective E2-transitions, which from spin I will populate a single rotational state at spin $I - 2$. In keeping with this discussion we show in Fig. 11.4, a schematic division of the rotational decay spectrum in the (E^*, I)-plane, in terms of discrete (a), damped (b) and compound (c) rotational transitions, calculated making use of the estimates given in Eq. (5.1), (9.13) and (11.38).

Examples like the ones discussed above testify to the subtle nature of chaos in many-body quantal systems like the atomic nucleus. They remind us also that damping of collective motion, like rotational motion, is, quite unrelated to chaotic motion.

The perspectives of studying the coexistence of collectivity and chaos in nuclei through studies of the rotational strength functions were pointed out by Guhr and Weidenmüller (1989). This question was addressed in the cranking model with random two-body interactions by Åberg (1990, 1992), and with more realistic interactions by Matsuo et al. (1993). The statistical properties of level spacings with regard to ordered or chaotic behaviour has been taken up by Garrett et al. (1991).

11.4 Two-Band Model of Rotational Damping: A Primer

In what follows we discuss the role the spread in rotational frequencies plays in the rotational pattern, within a two band model (cf. Fig. 11.5 and Lauritzen (1986)). The members of the two bands at spin I and $I - 2$ are denoted by $|a\rangle$, $|b\rangle$ and $|a'\rangle$, $|b'\rangle$, respectively. One assumes that they have identical quadrupole moments, that is,

$$\begin{cases} \langle a'|M(E2)|a\rangle = & \langle b'|M(E2)|b\rangle = & M_{E2}, \\ \langle a'|M(E2)|b\rangle = & \langle b'|M(E2)|a\rangle = & 0. \end{cases} \qquad (11.11)$$

The basis states interact with a coupling V, leading to the diagonalized states at spin I,

$$\begin{aligned} |\alpha\rangle &= X_a^\alpha|a\rangle + X_b^\alpha|b\rangle, \\ |\beta\rangle &= X_a^\beta|a\rangle + X_b^\beta|b\rangle, \end{aligned} \qquad (11.12)$$

and similarly at spin $I - 2$ leading to states $|\alpha'\rangle$, $|\beta'\rangle$. The value of the matrix element

$$(S_{a,a'})^{\frac{1}{2}} = \langle a'|M(E2)|a\rangle = A^{\frac{1}{2}}M_{E2}, \ A^{\frac{1}{2}} = X_{a'}^{\alpha'}X_a^\alpha + X_{b'}^{\alpha'}X_b^\alpha, \quad (11.13)$$

identifies the extent to which the mixed rotational bands retain the original pattern.

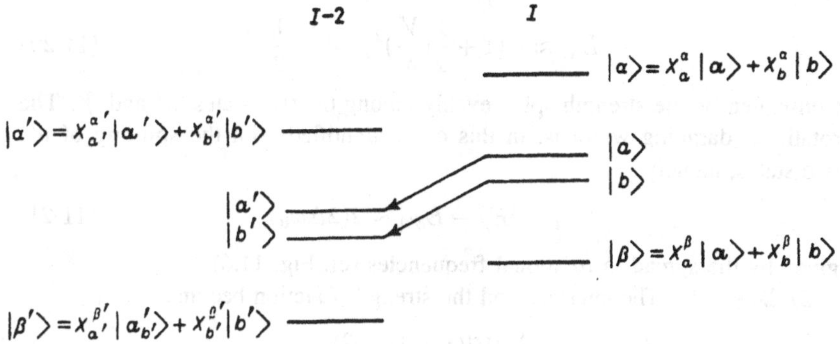

$$I-2 \qquad\qquad I$$

$$\underline{\qquad\qquad}\quad |a\rangle = x^a_a\,|a\rangle + x^a_b\,|b\rangle$$

$$|a'\rangle = x^{a'}_{a'}\,|a'\rangle + x^{a'}_{b'}\,|b'\rangle \quad \underline{\qquad\qquad}$$

$$\underline{\qquad\qquad}\quad |a\rangle$$
$$\underline{\qquad\qquad}\quad |b\rangle$$

$$\left.\begin{array}{l} |a'\rangle \\ |b'\rangle \end{array}\right\}\ \underline{\qquad\qquad}$$

$$\underline{\qquad\qquad}\quad |\beta\rangle = x^\beta_a\,|a\rangle + x^\beta_b\,|b\rangle$$

$$|\beta'\rangle = x^{\beta'}_{a'}\,|a'_b\rangle + x^{\beta'}_{b'}\,|b'\rangle \quad \underline{\qquad\qquad}$$

Figure 11.5. Schematic representation of the result of the diagonalization of the two band model.

Defining

$$\begin{cases} \varepsilon_a = \delta, & \varepsilon_b = -\delta, \\ \varepsilon_{a'} = \delta', & \varepsilon_{b'} = -\delta', \end{cases} \tag{11.14}$$

the Hamiltonian equation at spin I may be written as

$$\begin{pmatrix} \delta & V \\ V & \delta \end{pmatrix} \begin{pmatrix} X_a \\ X_b \end{pmatrix} = E \begin{pmatrix} X_a \\ X_b \end{pmatrix}, \tag{11.15}$$

and correspondingly at spin $I - 2$. The two solutions are given by

$$E_\alpha = -E_\beta = \delta(1 + (\frac{V}{\delta})^2)^{\frac{1}{2}}, \tag{11.16}$$

$$X^\alpha_b = X^\beta_b = [\frac{E_\alpha + \delta}{2E_\alpha}]^{\frac{1}{2}}, \quad X^\alpha_b = X^\beta_b = [\frac{E_\alpha - \delta}{2E_a}]^{\frac{1}{2}}. \tag{11.17}$$

The strength function is then given by

$$A = \frac{E_\alpha E_{\alpha'} + \delta\delta' + V^2}{2E_\alpha E_{\alpha'}}. \tag{11.18}$$

The quantities δ, δ' may be related to the spread in frequencies

$$(2\Delta\omega_0)^2 = \langle(2\omega_\mu)^2\rangle - \langle 2\omega_\mu\rangle^2 = (\delta - \delta')^2. \tag{11.19}$$

In what follows, we shall discuss the different regimes, characterized by the relative magnitudes of δ, $\Delta\omega_0$ and V. Both in the case of non-interacting bands ($V = 0$), and in the case of parallel bands ($\Delta\omega_0 = 0$), the rotational pattern is preserved, that is $A = 1$. In order to describe the region of high level density, we will make the assumption $\delta^2 \ll V^2$, $(2\Delta\omega_0)^2$, and study to extreme situations:

1) $\Delta\omega_0 > V$. In this case the energies and the strength function are given by

$$E_{\alpha'} \approx \delta'(1 + \frac{1}{2}(\frac{V}{\delta'})^2), \quad A \sim \frac{1}{2}. \tag{11.20}$$

Consequently, the strength splits evenly among the two states, α' and β'. The rotational damping width is, in this case, identified with the splitting of the two states, namely

$$\Gamma_{rot} \sim |E_\beta - E_{\beta'}| \sim 2(2\Delta\omega_0), \tag{11.21}$$

given by the spread in rotational frequencies (cf. Fig. 11.6).

2) $\Delta\omega_0 < V$. The energies and the strength function become

$$\begin{cases} E_\alpha \approx & \frac{\delta'}{|\delta'|}|V|(1 + \frac{1}{2}(\frac{\delta'}{V})^2), \\ A \approx & \frac{1}{2}(1 - \frac{\delta}{|\delta'|}) = \begin{cases} 0 & (\delta' > 0), \\ 1 & (\delta' < 0). \end{cases} \end{cases} \tag{11.22}$$

The rotational pattern is consequently retained, the rotational strength to spin $I-2$ being associated with the state at positive energy. However, the energy of this state is shifted relative to the energy of a perfect rotational band, $E = E_\alpha$, and we can relate the energy shift to half the rotational-damping width:

$$\Gamma_{rot} \sim 2|E_{\alpha'} - E_\alpha| \sim \frac{(\delta')^2}{V} \sim 2\frac{(2\Delta\omega_0)^2}{\Gamma_\mu}. \tag{11.23}$$

The damping width, $\Gamma_\mu \sim 2V$, is identified with the energy separation of the states $|\alpha\rangle$, $|\beta\rangle$ (cf. Fig. 11.7).

11.5 General Formulation of Rotational Damping

A compound state of a strongly rotating nucleus can be written as a linear combination

$$|\alpha(I)\rangle = \sum_\mu X_\mu^\alpha(I)|\mu(I)\rangle \tag{11.24}$$

of unperturbed rotational bands $|\mu(I)\rangle$, eigenstates of the cranked shell model Hamiltonian. A rotational partner to the state (11.24), defined according to

$$|\alpha(I-2)\rangle_{rot} = \sum_\mu X_\mu^\alpha|\mu(I-2)\rangle, \tag{11.25}$$

is connected by the collective quadrupole matrix element

$$\begin{aligned} & _{rot}\langle\alpha(I-2)|\mathcal{M}(E2)|\alpha(I)\rangle \\ = & \sum_\mu [X_\mu^\alpha(I)]^2 \langle\mu(I-2)|\mathcal{M}(E2)|\mu(I)\rangle \\ = & (2I+1)^{\frac{1}{2}}\overline{M_{E2}(\mu)_\alpha}, \end{aligned} \tag{11.26}$$

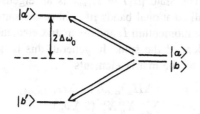

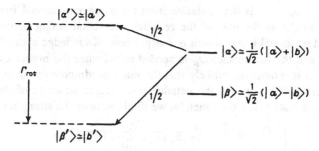

Figure 11.6. Two-band model of rotational damping for the case where the unperturbed bands display a dispersion in rotational frequency measured by $2\Delta\omega_0$ and couple through matrix elements $V < 2\Delta\omega_0$.

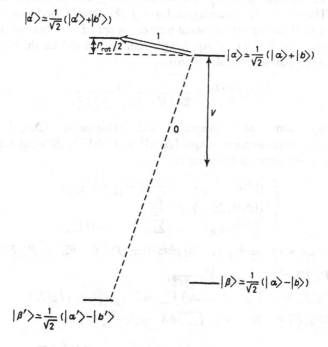

Figure 11.7. Same as Fig. 11.6 but now for $V > 2\Delta\omega_0$.

with the state $|\alpha(I)\rangle$. The state $|\alpha(I-2)\rangle_{rot}$ is an eigenstate of the total Hamiltonian, provided all rotational bands $\mu(I)$ have the same rotational frequency at a given angular momentum I, and the matrix elements of the residual interaction are independent of the spin. In general, this is not the case, and the states are connected by the matrix elements

$$\begin{cases} (\Delta H)_{\alpha,\alpha'} = & (\Delta H_\omega)_{\alpha,\alpha'} + (\Delta H_V)_{\alpha,\alpha'}, \\ (\Delta H_\omega)_{\alpha,\alpha'} = & \sum_\mu X_\mu^\alpha X_\mu^{\alpha'} (2\Delta\omega_\mu), \\ (\Delta H_V)_{\alpha,\alpha'} = & \sum_\mu X_\mu^\alpha X_\mu^{\alpha'} (\Delta V_{\mu,\mu'}). \end{cases} \qquad (11.27)$$

Here $\Delta\omega_\mu = \omega_\mu - \omega_0$ is the deviation from the average rotational frequency ω_0, while $\Delta V_{\mu,\mu'}$ is the part of the residual interaction acting between the unperturbed rotational bands which is I-dependent. Knowledge about the amplitudes X and the frequencies ω_μ is needed to calculate the matrix elements $(\Delta H)_{\alpha,\alpha'}$. It is, however, unlikely that the main quadrupole electromagnetic decay pattern will depend on the detailed microscopic structure of the compound nucleus state $|\alpha\rangle$. Consequently, we shall calculate the strength function

$$\begin{cases} P_\alpha(E) = & \frac{1}{2\pi} \frac{\Gamma_\alpha + \Delta}{(E_\alpha + \delta E_\alpha - E)^2 + \frac{1}{4}(\Delta + \Gamma_\alpha)^2}, \\ \Gamma_\alpha(E) = & \Delta \sum_\beta \frac{(\Delta H)_{\alpha,\beta}^2}{(E - E_\beta)^2 + \frac{1}{4}\Delta^2}, \\ \delta E_\alpha(E) = & \sum_\beta \frac{(\Delta H)_{\alpha,\beta}^2 (E - E_\beta)}{(E - E_\beta)^2 + \frac{1}{4}\Delta^2}, \end{cases} \qquad (11.28)$$

in terms of the amplitudes X_μ^α and frequencies ω_μ derived from a statistical model. The states $|\beta\rangle$ appearing in Eq. (11.28) are obtained diagonalizing the matrix ΔH among all rotational partners at spin $I-2$, but the one state $|\alpha(I-2)\rangle_{rot}$ (cf. Fig. 11.8). Applying a statistical model for the amplitudes leading to (cf. Lauritzen et al. (1986))

$$\varrho\langle(X_\mu^\alpha)^2\rangle = \frac{1}{2\pi} \frac{\Gamma_\mu}{(E - E_\mu)^2 + \frac{1}{4}\Gamma_\mu^2}, \qquad (11.29)$$

exchanging sums with integrals, and substituting $(\Delta\omega_\mu)^2$ and $(\Delta V_{\mu,\mu'})^2$ by their average values $(\Delta\omega_0)^2$ and $(\Delta V)^2$, different matrix elements can be worked out leading to

$$\begin{cases} \langle(\Delta H_\omega)_{\alpha,\alpha'}^2\rangle = & \frac{(2\Delta\omega_0)^2}{\pi\varrho} \frac{\Gamma_\mu}{(E_\alpha - E)^2 + \Gamma_\mu^2}, \\ \langle(\Delta H_V)_{\alpha,\alpha'}^2\rangle = & \frac{\Gamma_Q}{2\pi\varrho}, \\ \langle(\Delta H)_{\beta,\alpha}^2\rangle = & \sum_{\alpha'} \frac{1}{\varrho} P_{\alpha'} (\Delta H)_{\alpha,\alpha'}^2. \end{cases} \qquad (11.30)$$

One can now work out Eq. (11.28) obtaining ($P_0(E - E_\alpha) = P_\alpha(E)$)

$$\begin{cases} P_0(E) = & \frac{1}{2\pi} \frac{\Gamma_0}{(\delta E_0 - E)^2 + \frac{1}{4}\Gamma_0^2}, \\ \Gamma_0(E) = & \Gamma_Q + 2(2\Delta\omega_0) \int_{-\infty}^{+\infty} dX \frac{\Gamma_\mu}{(E - X)^2 + \Gamma_\mu^2} P_0(X), \\ \delta E_0(E) = & 2(2\Delta\omega_0)^2 \int_{-\infty}^{+\infty} dX \frac{E - X}{(E - X)^2 + \Gamma_\mu^2} P_0(X). \end{cases} \qquad (11.31)$$

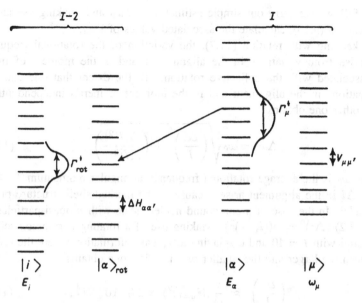

Figure 11.8. Illustration of the various steps in the calculation of the rotational-damping width Γ_{rot}. First, the basis band states $|\mu(I)\rangle$ are mixed by the residual interaction $V_{\mu\mu'}$ into the compound-nucleus eigenstates $|\alpha(I)\rangle$. Second, these states decay sharply to their "rotational partners" at spin $I - 2$, $|\alpha(I - 2)\rangle_{rot}$, and finally the rotational partner states are fragmented among the energy eigenstates $|i(I - 2)\rangle$ at spin $I - 2$ by the coupling $\Delta H_{\alpha\alpha'}$. The properties ω_μ, E_α, E_i of the different states are indicated at the bottom of the figure. (From Lauritzen et al. (1986).)

11.5.1 Simple estimates

The relations given in Eqs. (11.31) determine the probability of finding the state $|\alpha(I - 2)\rangle_{rot}$ per unit energy in a self-consistent way. They have simple analytic solutions in the case when Γ_μ is much smaller and much larger than $2\Delta\omega_0$. One then obtains the width of the strength function

$$\Gamma_{rot} = \begin{cases} 2(2\hbar\Delta\omega_0), & \Gamma_Q, \Gamma_\mu < 2\Delta\omega_0, \\ \Gamma_Q + \frac{2(2\hbar\Delta\omega_0)^2}{\Gamma_\mu}, & \Gamma_\mu \gg 2\Delta\omega_0. \end{cases} \quad (11.32)$$

Consequently, the rotational damping is controlled by the spreading in rotational frequencies of the unperturbed rotational bands and by the time $\frac{\hbar}{\Gamma_\mu}$ the state (11.24) of the compound nucleus spends in each of these bands. When this time is long compared with the spread in frequencies (i.e., $\frac{\hbar}{\Gamma_\mu} > (2\Delta\omega_0)^{-1}$), Γ_{rot} is expected to increase with temperature. Otherwise Γ_{rot} becomes a decreasing function of T. In fact, if the nucleus rotating with a given energy and angular momentum changes its intrinsic state only after a fraction of revolution and before it has had time to experience the differential change in the angular velocity, the expected spread in transition energies $2\Delta\omega_0$ is averaged out. In what follows we carry out simple estimates of $\Delta\omega_0$ and, making use also of the relation (9.13), calculate the associated values of Γ_{rot}.

In keeping with relation (11.7), the variation of the rotational frequency can arise from variations in the alignment i and in the moment of inertia $\mathcal{J}$ associated with the collective rotation. To the extent that one can treat fluctuations in the alignment and in the moment of inertia independently of each other one obtains

$$\Delta\omega_0 = \omega_0\sqrt{\left(\frac{\Delta i}{I}\right)^2 + \left(\frac{\Delta\mathcal{J}^{(2)}}{\mathcal{J}^{(2)}}\right)^2}, \tag{11.33}$$

where ω_0 is the average rotational frequency at angular momentum I. Variations Δi in the alignment arise because of the marked shell structure present in nuclei. In the case of a compound nuclear state with v quasi-particles (cf. Sect. 7.2) $(\Delta i)^2 = v\langle(i_{v=1})^2\rangle$. Making use of a rotating harmonic-oscillator potential with $I = 40$ and displaying a typical deformation where the relation between the larger and the smaller axis is 1.3:1, one obtains

$$\left(\frac{\Delta i}{I}\right)^2 \approx \frac{v}{I^2}\langle(i_{v=1})^2\rangle \approx 2.8\cdot 10^{-3}U^{\frac{1}{2}}, \tag{11.34}$$

where U is the heat energy of the system, that is the energy above the yrast line (cf. Eq. (5.12)). Fluctuations in the moment of inertia are connected with fluctuations in the nuclear shape. At finite temperature, the system can be at any point of the (β, γ)-plane with a probability given by

$$P = \frac{\exp[-F(\beta, \gamma; I, T)/T]}{Z}, \tag{11.35}$$

where $F(\beta, \gamma; I, T)$ is the free energy, and Z the partition function.

Making the ansatz that the free energy depends quadratically on β and using the rotating-liquid-drop model and the parametrization (Hill and Wheeler (1953))

$$\mathcal{J}^{(2)} = \mathcal{J}_{sp}(1 + (\frac{5}{4\pi})^{\frac{1}{2}}\cos(\gamma - \frac{\pi}{3})) \tag{11.36}$$

of the moment of inertia in terms of the parameters β and γ one obtains

$$\left(\frac{\Delta\mathcal{J}^{(2)}}{\mathcal{J}^{(2)}}\right)^2 = \frac{5}{4\pi}\frac{T}{C_2} \approx 1.4\cdot 10^{-3}U^{\frac{1}{2}}. \tag{11.37}$$

The quantity C_2 is the quadrupole restoring force in the liquid-drop model, $C_2 \sim 70$ MeV (Bohr and Mottelson (1975)).

Collecting the different terms together, one obtains

$$\Delta\omega_0 \approx 36U_{\mathrm{MeV}}^{\frac{1}{4}} \text{ keV}, \tag{11.38}$$

where the typical values $I = 40$ and $\frac{\mathcal{J}^{(2)}}{\hbar^2} \sim 70$ MeV^{-1} were used.

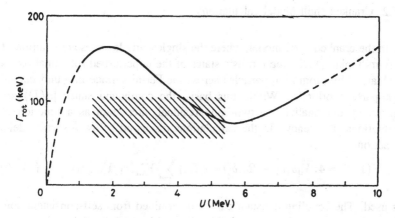

Figure 11.9. The rotational damping width described by Eq. (11.39) as a function of the excitation energy U above the yrast line. The experimental information (Garret et al. (1986), Bracco et al. (1996)) is represented by a shaded area.

We can now calculate the rotational-damping width (11.32) obtaining

$$\Gamma_{rot} = \begin{cases} 0.14U^{\frac{1}{4}} \text{ (MeV)}, & U_c < 2.5 \text{ MeV}, \\ 0.07U^{-\frac{1}{2}} \text{ (MeV)}, & U_c > 2.5 \text{ MeV}. \end{cases} \tag{11.39}$$

The threshold heat energy U_c for which motional narrowing is expected was obtained setting Γ_μ equal to $2(2\hbar\Delta\omega_0)$.

In expression (11.39) only the frequency term in the coupling ΔH_ω of the Eq. (11.30) has been considered. The remaining width, Γ_Q, arising from the spin dependence of the residual interaction H_V, can be related to the damping width Γ_μ according to

$$\Gamma_Q \approx 2\pi\varrho_\mu(\Delta V)^2 \approx \Gamma_\mu \left(\frac{\Delta H}{V}\right)^2. \tag{11.40}$$

In a cranked uniform model, the quantity $(\frac{\Delta H_V}{V})^2$ is estimated to be $\sim 10^{-2}$. Making use of (9.13) one can thus write

$$\Gamma_Q \sim 5U \text{ (keV)}. \tag{11.41}$$

This damping mechanism will control Γ_{rot} for an excitation energy U_c such that

$$\Gamma_{rot}(U_c') = 0.07(U_c')^{-\frac{1}{2}} = 0.5 \cdot 10^{-2}U_c', \tag{11.42}$$

leading to $U_c' \sim 6$ MeV. The above results are shown in Fig. 11.9. The experimental data are consistent with a value of $100 \text{ keV} \leq \Gamma_{rot} \leq 200 \text{ keV}$, in overall agreement with the calculation.

11.5.2 Cranked shell model calculations

Within the cranked shell model, where the single-particle states are solution of the Hamiltonian (11.2), the intrinsic states of the unperturbed rotational bands are obtained by promoting particles across the Fermi surface into unoccupied single-particle orbitals. Within this basis, the compound states $|\alpha(I)\rangle$ are obtained by diagonalizing a two-body residual interaction as a function of the rotational frequency. In the calculations discussed below a surface-delta interaction

$$V(1,2) = 4\pi V_0 \delta(r_1 - R_0)\delta(r_2 - R_0) \sum_{\lambda\mu} Y^*_{\lambda\mu}(r_1) Y_{\lambda\mu}(r_2) \qquad (11.43)$$

was used. The coupling constant can be determined from self-consistent conditions relating the variations in the density and in the potential.

Calculations have been carried out (Matsuo et al. (1993), Matsuo et al. (1997), Broglia et al. (1996)) for the case of the nucleus ^{168}Yb and the unperturbed bands $|\mu(I)\rangle$ including 10^3 states at each spin I. The associated matrices of the interaction given in Eq. (11.43) have been diagonalized for each parity and signature as a function of the rotational frequency for a value of the coupling constant $V_0 = 27.5/A$ MeV. In this way one obtains the compound nuclear states (11.24). The truncation in the size of the basis set used has been checked to give stable results for the lowest 300 levels for each parity and signature. These states cover an interval in excitation energy of approximately 2.4 MeV above the yrast line. The stretched E2 transitions strength were calculated between levels at the rotational frequency ω and those at $\omega - 2/\mathcal{J}$ corresponding to $I - 2$, where $\mathcal{J}$ is the moment of inertia.

Results of the calculation for the levels with parity + and signature 0 are displayed in Fig. 11.10 as a function of the angular momentum (rotational frequency) in the interval $36 < I < 46$. If the quadrupole decay out of a compound nuclear state at spin I populates with a probability larger than 71% a single state at spin $I-2$, the two states are joined by a continuous line. If this probability is in the range between 71% and 51%, the levels are joined by a dashed line. The presence of lines joining a sequence of states is considerably larger in the energy region close to the yrast line.

In keeping with the picture of rotational damping, as the heat energy increases, the full continuous lines as well as the dashed lines in Fig. 11.10 become gradually shorter and shorter, connecting progressively fewer and fewer compound states belonging to a rotational band. Quadrupole transitions fan out reflecting the loss of correlation between different I-spin states. The damping process shows however conspicuous irregularities. In fact, even at excitation energies above the yrast line of the order of 2 MeV, where the distance between unperturbed rotational levels is of the order of kiloelectronvolts and where typical matrix elements between unperturbed bands are of the order of 20 keV, one finds continuous lines joining two, in some cases three states at

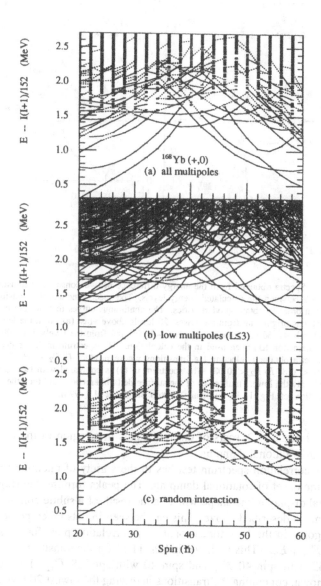

Figure 11.10. The compound energy levels of ^{168}Yb defined in Eq. (11.24), and calculated as described in the text are shown as a function of angular momentum, subtracting the rotational energy of a rigid rotor of moment of inertia $\mathcal{J} = 76\ \hbar^2\ \text{MeV}^{-1}$. Two levels are connected by a solid (dashed) line if the associated transition probability P is larger than 0.7 (lies in the interval 0.5–0.7). (a) refers to the calculation carried out diagonalizing the full surface delta interaction, while in (b) only the multipoles $\lambda \leq 3$ are considered. (c) refers to results of a calculation performed with a random force, whose average mean square root value of the matrix element was set equal to $\langle V_{\mu\mu'}^2 \rangle_{\lambda>3}^{1/2} \approx 12$ keV. (From Broglia et al. (1996).)

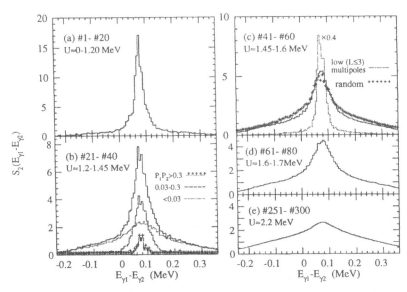

Figure 11.11. Spectra obtained from the energy difference of two consecutive transitions $I + 2 \rightarrow I \rightarrow I - 2$ between the calculated energy levels. The spectra include the transitions with $30 < I < 50$ in order to obtain good statistics. The continuous curves in (a–e) are constructed from the transitions originating from the lowest 20 levels above yrast (a), from the levels 21–40 (b), from the levels 41–60 (c), from the levels 61–80 (d) and from the levels 251–300 (e). The surface delta interaction has been used in the calculation. The spectrum in (b) is decomposed in three subspectra obtained selecting the transitions whose product $P_1 P_2$ is $P_1 P_2 > 0.3$, $0.03 < P_1 P_2 < 0.3$, $P_1 P_2 < 0.03$. The spectrum in (c) is compared with two spectra, one obtained using only the lowest multipoles of the surface-delta interaction, and the other using the random interaction (cf. Fig. 11.10 for details). From Broglia et al. (1996).

spin $I + 2$, I and $I - 2$, scars of discrete bands in a region of the U–I plane controlled by rotational damping.

The γ–γ correlation spectrum testifies to the variety of phenomena associated with the onset of rotational damping. The peaks forming the ridge along the diagonal display a line shape which is the result of a subtle combination of the phenomena described above. This can be seen in cuts taken perpendicular to the diagonal in the two-dimensional γ–γ correlation plot, for a variety of values of $E_{\gamma_1} + E_{\gamma_2}$. This is shown in Fig. 11.11 for the transitions connecting the spin 42 with spin 40 E_{γ_1} and spin 40 with spin 38 E_{γ_2}. Four different bins are shown, containing the transitions involving the lowest 20 levels at spin 40, and from the levels 21–40, 41–60, 61–80 and 251–300, corresponding to the energy intervals $U = 0$–1.2 MeV, 1.2–1.45 MeV, 1.45–1.6 MeV, 1.6–1.7 MeV, and 2.2–2.25 MeV, respectively. A line shape made out of the sum of line widths displaying different widths is apparent. This can be better seen by displaying also the line shapes including transitions such that the product of the probability of the first and second transition is larger than 0.5, 0.3, 0.05 and 0.01 respectively.

The very narrow component of the strength function displays a width of the order of 20 keV and can be correlated with the distribution of values of the moment of inertia associated with well-defined, near-yrast states (continuous lines in Fig. 11.10). The next broader component (0.1–$0.3 < P_1P_2 < 0.5$), displaying a width between 50 and 80 keV (cf. Fig. 11.11(a)), depending on the bin under question, can be connected with coincidence events arising from the scars of rotational bands at relatively high excitation energy (short traits connecting 3–4 states in Fig. 11.10). The width of this component changes with the energy interval covered by the bin, and is connected essentially to the compound nucleus damping width Γ_μ. Finally, a broad structure is observed $P_1P_2 < 0.01$) with a width of the order of 200–250 keV. This width arises from the decay of states which from spin I populate a large number of states at spin $I - 2$, and can be associated with the "real" rotational damping width. As the energy is progressively increased ($U > 2$ MeV) this is essentially the only component that survives. The onset of rotational damping is thus not expected to be a sharp event taking place at a given heat energy, but to proceed in steps over a broad heat energy region (1–1.5 MeV).

References

Åberg, S. (1990). *Phys. Rev. Lett.* **64**:3119.

Åberg, S. (1992). *Progr. Part. Nucl. Phys.* **28**:11.

Åberg, S. (1996). In: *Workshop on Gammasphere Physiscs*, ed. M.A. Delaplanque. Singapore: World Scientific.

Alhassid, Y. (1994). *Nucl. Phys.* **A569**:37c.

Alhassid, Y., and Bush, B. (1989). *Phys. Rev. Lett.* **63**:2452.

Alhassid, Y., and Bush, B. (1990). *Nucl. Phys.* **A509**:461.

Alhassid, Y., and Bush, B. (1990a). *Phys. Rev. Lett.* **65**:2527.

Alhassid, Y., and Zingman, J. (1984). *Phys. Rev.* **C30**:684.

Alhassid, Y., Bush, B., and Levit, S. (1988). *Phys. Rev. Lett.* **61**:294.

Arve, P., Bertsch, G.F., Lauritzen, B., and Puddu, G. (1988). *Ann. Phys. (N.Y.)* **183**:309.

Bardeen, J., Cooper, L.N., and Schieffer, J.R. (1957). *Phys. Rev.* **106**:162, 1175.

Barranco, F., and Broglia, R.A. (1985). *Phys. Lett.* **B151**:90.

Barranco, F., and Broglia, R.A. (1987). *Phys. Rev. Lett.* **59**:2724.

Barranco, F., Bertsch, G.F., Broglia, R.A., and Vigezzi, E. (1992). *Nucl. Phys.* **A512**:253.

Bauer, M., Hernandez-Saldana, E., Hodgson, P., and Quintillana, J. (1982). *J. Phys.* **G8**:525.

Beene, J.R., Varner, R.L., and Bertrand, F.E. (1988). *Nucl. Phys.* **A482**:407c.

Beene, J.R., et al. (1989). *Phys. Rev.* **C39**:1307.

Beene, J.R., et al. (1990). *Phys. Rev.* **C41**:920.

Beene, J.R., et al. (1994). *Nucl. Phys.* **A569**:163c.

Behr, J.A., et al. (1993). *Phys. Rev. Lett.* **70**:3201.

Berestetskii, V.B., Lifshitz, E.M., and Pitaevskii (1976). *Relativistic Quantum Theory*. New York: Pergamon Press.

Berman, B.L., and Fultz, S.C. (1975). *Rev. Mod. Phys.* **47**:713.

Bertrand, F.E. (1976). *Ann. Rev. Nucl. Sci.* **26**:457.

Bertrand, F.E., Beene, J.R., and Sjoreen, T.P. (1984). *J. Phys. (Paris)* **45**:C4-99.

Bertrand, F.E., Beene, J.R., and Horen, D.J. (1988). *Nucl. Phys.* **A482**:287c.

Bertsch, G.F. (1983). *Scient. Am.* **248**:62.

Bertsch, G.F. (1988). *Physics Today*, June.

Bertsch, G.F., and Broglia, R.A. (1986). *Physics Today*, August.

Bertsch, G.F., and Broglia, R.A. (1994). *Oscillations in Finite Quantum Systems*. Cambridge: Cambridge University Press.

Bertsch, G.F., and Siemens, P.J. (1983). *Phys. Lett.* **B126**:9.

Bertsch, G.F., and Tsai, S.F. (1975). *Phys. Rep.* **18**:125.

Bertsch, G.F., Bortignon, P.F., Broglia, R.A., and Dasso, C.H. (1979). *Phys. Lett.* **80B**:161.

Bertsch, G.F., Bortignon, P.F., and Broglia, R.A. (1983). *Rev. Mod. Phys.* **55**:287.

Bes, D.R., and Sorensen, R. (1969). *Adv. Nucl. Phys.* **2**:129.

Bjørnholm, S., and Lynn, J.E. (1980). *Rev. Mod. Phys.* **52**:725.

Blaizot, J.P., Berger, J.F., Decharge J., and Girod, M. (1995). *Nucl. Phys.* **A591**:435.

Block, B., and Feshbach, H. (1963). *Ann. Phys. (N.Y.)* **23**:47.

Bogoliubov, N.N. (1968). *Lectures on Quantum Statistics*, Vol. 1. London: MacDonald Technical and Scientific.

Bohr, N. (1936). *Nature* **137**:344.

Bohr, A., and Mottelson, B.R. (1969). *Nuclear Structure*, Vol. 1 New York: Benjamin.

Bohr, A., and Mottelson, B.R. (1975). *Nuclear Structure*, Vol. 2. New York: Benjamin.

Bohr, A., and Mottelson, B.R. (1981). *Phys. Scr.* **24**:71.

Bohr, A., Mottelson, B.R., and Pines, D. (1967). *Phys. Rev.* **110**:936.

Bohr, N., and Wheeler, J.A. (1939). *Phys. Rev.* **56**:426.

Bortignon, P.F., and Broglia, R.A. (1981). *Nucl. Phys.* **A371**:405.

Bortignon, P.F., Broglia, R.A., and Xia Ke-Ding (1984). *J. Phys. (Paris)* **45**:C4-209.

Bortignon, P.F., Broglia, R.A., and Bertsch, G. F. (1984a). *Phys. Lett.* **148B**:20.

Bortignon, P.F., Broglia, R.A., Bes, D.R., and Liotta, R. (1977). *Phys. Rep.* **30C**:305.

Bortignon, P.F., Broglia, R.A., Bertsch, G.F., and Pacheco, J. (1986). *Nucl. Phys.* **A460**:149.

Bortignon, P.F., Bracco, A., Brink, D., and Broglia, R.A. (1991). *Phys. Rev. Lett.* **67**:3360.

Bortignon, P.F., Braguti M., Brink, D.M., Broglia, R.A., Brusati, C., Camera, F., Cassing, W., Cavinato, M., Giovanardi, N., Gulminelli, F. (1995). *Nucl. Phys.* **A583**:101c.

Bracco, A., et al. (1988). *Phys. Rev. Lett.* **60**:2603.

Bracco, A., et al. (1989). *Phys. Rev.* **C39**:725.

Bracco, A., et al. (1989a). *Phys. Rev. Lett.* **62**:2080.

Bracco, A., Bortignon, P.F., and Broglia, R.A. (1992). In *Nuclear Collisions, from the Mean Field into Fragmentation Regime*, eds. Detraz and Kienle, p. 119. Amsterdam: North-Holland.

Bracco, A., Camera, F., Mattiuzzi, M., Million, B., Pignanelli, M., Gaardhøje, J.J., Maj, A., Ramsøy, T., Tveter, T., and Zelazny, Z. (1995). *Phys. Rev. Lett.* **74**:3748.

Bracco, A., Bosetti, P., Frattini, S., Vigezzi, E., Leoni, S., Herskind, B., Døssing, T., and Matsuo, M. (1996). *Phys. Rev. Lett.* **76**:4484.

Brack, M., and Quentin, P. (1974). *Phys. Lett.* **B52**:159.

Brack, M., and Quentin, P. (1974a). *Physica Scripta* **10A**:163.

Brack, M., Damgaard, J., Jensen, A.S., Pauli, H.C., Strutinsky, V.M., and Wong, C. Y. (1972). *Rev. Mod. Phys.* **44**:320.

Brandenburg, S., et al. (1989). *Phys. Rev.* **C39**:2448.

Brink, D.M. (1955). Ph.D. Thesis, Oxford University (unpublished).

Broglia, R.A. (1988). In *Frontiers and Borderlines in Many-Particle Physics*, eds. Broglia and Schrieffer, p. 204. Amsterdam: North-Holland.

Broglia, R.A., and Winther, A. (1991). *Heavy Ion Reactions*. Reading: Addison-Wesley.

Broglia, R.A., Hansen, O., and Riedel, C. (1973). *Adv. Nucl. Phys.* Vol. 6, p. 287. New York: Plenum.

Broglia, R.A., Døssing, T., Lauritzen, B., and Mottelson, B.R. (1987). *Phys. Rev. Lett.* **58**:326.

Broglia, R.A., Ormand, W.E., and Borromeo, M. (1988). *Nucl. Phys.* **A482**:141c.

Broglia, R.A., Barranco, F., and Vigezzi, E. (1993). *Jap. J. Appl. Phys.* Ser. **9**:164.

Broglia, R.A., Barranco, F., Bertsch, G.F., and Vigezzi, E. (1994). *Phys. Rev.* **C49**:552.

Broglia, R.A., Døssing, T., Matsuo, M., Vigezzi, E., Bosetti, P., Bracco, A., Frattini, S., Heskind, B., Leoni, S., and Rasmussen, P. (1996). *Z. Phys.* **A356**:259.

Broglia, R.A., Døssing, T., Mottelson, B.R., Vigezzi, E. (1998). To be published.

Brown, G.E., and Osnes, E. (1985). *Phys. Lett.* **B159**:223.

Brun, R., et al. (1986). CERN/DD/EE/84-1.

Buda, A., et al. (1995). *Phys. Rev. Lett.* **75**:798.

Butsch, R., et al. (1991). *Phys. Rev.* **C44**:1515.

Caldwell, J.T., et al. (1980). *Phys. Rev.* **C21**:1215.

Camera, F. (1992). Ph.D. Thesis, University of Milano (unpublished).

Camera, F. (1995). ENEA/NSC/COC(95)1, 155.

Carlos, P., et al. (1974). *Nucl. Phys.* **A219**:61.

Chakrabarty, D.R., et al. (1987). *Phys. Rev.* **C36**:1886.

Chandrasekhar, S. (1969). *Ellipsoidal Figures of Equilibrium.* New Haven: Yale University Press.

Chomaz, Ph. (1995a). *Phys. Lett.* **B347**:1.

Chomaz, Ph., and Frascaria, N. (1995). *Phys. Rep.* **252**:275.

Chomaz, Ph., et al. (1993). *Nucl. Phys.* **A563**:509.

Civitarese, O., Broglia, R.A., and Dasso, C.H. (1984). *Ann. Phys.* **156**:142.

Cohen, B.L. (1965). *Am. J. Phys.* **33**:1011.

Cohen, B.L. (1973). *Concepts in Nuclear Physics.* New York: McGraw-Hill.

Cohen, S., Plasil, F., and Swiatecki, W. (1974). *Ann. Phys. (N.Y.)* **22**:406.

Colò, G., Bortignon, P.F., Nguyen, van Giai, Bracco, A., and Broglia, R.A. (1992). *Phys. Lett.* **B276**:279.

Colò, G., Nguyen, van Giai, Bortignon, P.F., and Broglia, R.A. (1994). *Phys. Rev.* **C50**:1496.

Danos, M. (1958). *Nucl. Phys.* **5**:23.

Dattagupta, S. (1987). *Relaxation Phenomena in Condensed Matter Physics.* New York: Academic Press.

Dean, D.J., et al. (1994). *Phys. Rev. Lett.* **72**:4066.

De Blasio, F.V., Cassing, W., Tohyama, M., Bortignon, P.F., and Broglia, R.A. (1992). *Phys. Rev. Lett.* **68**:1663.

De Gennes, P.G. (1994). *Les Objects Fragiles.* Paris: Plon.

De Leo, R., Brandenburg, S., Drentje, A.G., Harakeh, M.N., Janszen, H., and Van Der Woude, A. (1985). *Nucl. Phys.* **A441**:591.

Dias, H., et al. (1986). *Phys. Rev. Lett.* **57**:1998.

Dilg, W., Schantl, W., Vonach, H., Uhl, M. (1973). *Nucl. Phys.* **A217**:269.

Dirac, P.A.M. (1935). *The Principles of Quantum Mechanics.* Oxford: Claredon Press.

Donati, P., Pizzochero, P.M., Bortignon, P.F., and Broglia, R.A. (1994). *Phys. Rev. Lett.* **72**:2835.

Donati, P., Bortignon, P.F., and Broglia, R.A. (1996). *Z. Phys.* **A354**:249.

Donati, P., Giovanardi, N., Bortignon, P.F., and Broglia, R.A. (1996a). *Phys. Lett.* **B383**:15.

Døssing, T., Herskind, B., Leoni, S., Bracco, A., Broglia, R.A., Matsuo, M., and Vigezzi, E. (1996). *Phys. Rep.* **268**:1.

Draper, J.E., Dines, E.L., Delaplanque, M.A., Diamond, R.M., and Stephens, F.S. (1986). *Phys. Rev. Lett.* **50**:409.

Emling, H. (1994). *Prog. Part. Nucl. Phys.* **33**:729.

Erba, E., Facchini, U., and Saetta-Menichella, E. (1961). *Nuovo Cimento* **22**:1237.

Ericson, T. (1958). *Nucl. Phys.* **8**:265; **9**:697.

Esbensen, H., and Bertsch, G.F. (1983). *Phys. Rev.* **C28**:355.

Esbensen, H., and Bertsch, G.F. (1984). *Ann. Phys. (N.Y.)* **157**:255.

Esbensen, H., and Bertsch, G.F. (1984a). *Phys. Rev. Lett.* **52**:2257.

Eyrich, W., et al. (1983). *Lecture Notes Phys.* **158**:283.

Feshbach, H., Kerman, A., and Lemmer R. H. (1967). *Ann. Phys. (N.Y.)* **41**:230.

Feshbach, H., Kerman, A., and Koonin, S. (1980). *Ann. Phys. (N.Y.)* **125**:429.

Feynman, R.P., and Hibbs, A.R. (1965). *Quantum Mechanics and Path Integral.* New York: McGraw-Hill.

Gaardhøje, J.J. (1992). *Ann. Rev. Nucl. Part. Sci.* **42**:483.

Gaardhøje, J.J., et al. (1986). *Phys. Rev. Lett.* **56**:1783.

Gaardhøje, J.J., et al. (1987). *Phys. Rev. Lett.* **59**:1409.

Galès, S. (1980). *Proceedings of the Mazurian Summer School.* Mikolajki, Poland.

Galès, S., Crawley, G.M., Weber, D., and Zwieglinski, A. (1978). *Phys. Rev.* **C18**:2475.

Gallardo, M., Diebel, M., Døssing, T., and Broglia, R.A. (1985). *Nucl. Phys.* **A443**:415.

Garret, J.D., Hagemann, G., and Herskind, B. (1986). *Ann. Rev. Nucl. Sci.* **36**:419.

Garret, J.D., German, J.R., Courtney, L., and Espino, J.M. (1991). In *Future Directions in Nuclear Physiscs with 4π-Gamma Detection System of the New Generation*, eds. Dudek and Haas, p. 345. New York: American Institute of Physics.

Giannini, M.M, Ricco, G., and Zucchiatti, A. (1980). *Ann. Phys.* **124**:208.

Giovanardi, N. (1996). Ph.D. Thesis, University of Milan (unpublished).

Giovanardi, N., Bortignon, P.F., Broglia, R.A., and Huang, W. (1996). *Phys. Rev. Lett.* **77**:24.

Gonin, M., Cooke, L., Fornal, B., et al. (1989). *Nucl. Phys.* **A495**:139c.

Gosset, C.A., Snover, K.A., Behr, J.A., Feldman, G., and Osborne, J.L. (1985). *Phys. Rev. Lett.* **54**:1486.

Grangè, P., and Weidenmüller, H. (1980). *Phys. Lett.* **96B**:26.

Grodzins, L., et al. (1984). *Ann. Israel Phys. Soc.* **7**:227.

Groningen Conference (1995). *Proceedings of the Groningen Conference on Giant Resonances.* Groningen, The Netherlands. *Nucl. Phys.* **A599** (1996).

Guhr, T., and Weidenmüller, H. (1989). *Ann. Phys.* **193**:472.

Guillot, J., van de Wiele, J., Langevin-Joliot, H., Gerlic, E., Didelez, J.P., Duhamel., G., Perrin, G., Buenerd, M., and Chaovin, J. (1980). *Phys. Rev.* **C21**:879.

Gull Lake Conference (1993). *Proceedings of the Gull Lake Nuclear Physics Conference on Giant Resonances*, Gull Lake, Michigan. *Nucl. Phys.* **A569** (1994).

Gundlach, J.H., et al. (1990). *Phys. Rev. Lett.* **65**:2523.

Harakeh, M.N., Dowell, D.H., Feldman, G., Garman, E.F., Loveman, R., Osborne, J.L., and Snover, K.A. (1986). *Phys. Lett.* **B176**:297.

Hauser, W., and Feshbach, H. (1952). *Phys. Rev.* **87**:366.

Herskind, B. (1984). In *Nuclear Structure and Heavy-Ion Dynamics*, eds. Moretto and Ricci, p. 68. Amsterdam: North-Holland.

Herskind, B., Døssing, T., Leoni, S., Matsuo, M., and Vigezzi, E. (1992). *Prog. Part. Nucl. Phys.* **28**:235.

Hill, D.L., and Wheeler, J.A. (1953). *Phys. Rev.* **89**:1102.

Hilscher, D., and Rossner, H. (1992). *Ann. Phys. Fr.* **17**:471.

Hubbard, J. (1959). *Phys. Rev. Lett.* **3**:77.

Jacob, G., and Maris, T.A.J. (1973). *Rev. Mod. Phys.* **45**:6.

Kerman, A.K., and Levit, S. (1981). *Phys. Rev.* **C24**:1029.

Kerman, A.K., Levit, S., and Troudet, T. (1983). *Ann. Phys. (N.Y.)* **148**:436.

Kicińska-Habior, M., et al. (1993). *Phys. Lett.* **B308**:225.

Kilgus, G., Kühner, G., Muller, S., Richter, A., and Knüpfer, W. (1987). *Z. Phys.* **A326**:41.

Koonin, S.E., Dean, D.J., and Langanke, K. (1997). *Phys. Rep.* **278**:1.

Kramers, H.A. (1940). *Physica VII* **87**:366.

Kubo, R., and Nagamiya, T. (1969). *Solid State Physics*. New York: Mc-Graw Hill.

Kühner, G., et al. (1981). *Phys. Lett.* **B104**:189.

Landau, L.D. (1957). *Zh. Eksp. Teor. Fiz.* **32**:59.

Lauritzen, B. (1986). Ph.D. Thesis, University of Copenhagen (unpublished).

Lauritzen, B., Døssing, T., and Broglia, R.A. (1986). *Nucl. Phys.* **A457**:61.

Lauritzen, B., Døssing, T., Broglia, R.A., and Ormand, W.E. (1988). *Phys. Lett.* **B207**:238.

Lauritzen, B., Anselmino, A., Broglia, R.A., and Bortignon, P.F. (1993). *Ann. Phys.* **223**:216.

Lauritzen, B., Bortignon, P.F., Broglia, R.A., and Zelevinsky, V.G. (1995). *Phys. Rev. Lett.* **74**:5190.

Leander, G.A. (1982). *Phys. Rev.* **C25**:2780.

Le Faou, J.H., et al. (1994). *Phys. Rev. Lett.* **72**:3321.

Legnaro Conference (1987). *Proceedings of the First Topical Meeting on Giant Resonance Excitation in Heavy-Ion Collisions.* Legnaro, Italy. *Nucl. Phys.* **A482** (1988).

Levinger, J.S. (1960). *Nuclear Photo-Disintegration.* Oxford: Oxford University Press.

Levit, S. (1988). In *The Response of Nuclei Under Extreme Conditions*, eds. R.A. Broglia and G.F. Bertsch, p. 87. New York: Plenum.

Levit, S., and Alhassid, Y. (1984). *Nucl. Phys.* **A413**:439.

Liotta, R., and Sorensen, R. (1978). *Nucl. Phys.* **A279**:136.

Lipparini, E., and Stringari, S. (1989). *Phys. Rep.* **175**:103.

Mahan, G. (1981). *Many-Particle Physics.* New York: Plenum Press.

Mahaux, C., and Ngô, H. (1981). *Phys. Lett.* **B100**:285.

Mahaux, C., Bortignon, P.F., Broglia, R.A., and Dasso, C. (1985). *Phys. Rep.* **120**:1.

Maj, A., Gaardhøje, J.J., Atac, A., Mitarai, S., Nyberg, J., Virtanen, A., Bracco, A., Camera, F., Million, B., and Pignanelli, M. (1994). *Nucl. Phys.* **A571**:185.

Matsuo, M., Døssing, T., Vigezzi, E., and Broglia, R.A. (1993). *Phys. Rev. Lett.* **70**:2694.

Matsuo, M., Døssing, T., Vigezzi, E., Broglia, R.A., and Yoshida, K. (1997). *Nucl. Phys.* **A617**:1.

Mattiuzzi, M., Bracco, A., Camera, F., Million, B., Pignanelli, M., Gaardhøje, J.J., Maj, A., Ramsøy, T., Tveter, T., and Zelazny, Z. (1995). *Phys. Lett.* **B364**:13.

Mattiuzzi, M., Bracco, A., Camera, F., Ormand, W.E., Million, B., Pignanelli, M., Gaardhøje, J.J., Maj, A., Ramsøy, T., and Tveter, T. (1997). *Nucl. Phys.* **A612**:262.

Migdal, A.B. (1967). *Theory of Finite Fermi Systems and Applications to Atomic Nuclei.* New York: Interscience.

Migneco, E., et al. (1992). *Nucl. Instr. Meth.* **A314**:31.

Morel, P., and Nozières, P. (1962). *Phys. Rev.* **126**:1909.

Morsch, H.P., et al. (1982). *Phys. Lett.* **B119**:311.

Mottelson, B.R. (1962). In *Nuclear Spectroscopy*, ed. G. Racah, p. 44. New York: Academic Press.

Mottelson, B.R. (1993). In *The Frontiers of Nuclear Spectroscopy*, eds. Y. Yoshizawa, H. Kusakari and T. Otsuka, p. 7. Singapore: World Scientific.

Mottelson, B.R. (1993a). *Nucl. Phys.* **A557**:717c.

Nebbia, G., et al. (1986). *Phys. Lett.* **B176**:20.

Negele, J.W. (1982). *Rev. Mod. Phys.* **54**:913.

Newton, J.O., Herskind, B., Diamond, R.M., Dines, J.E., Draper, J.E., Lindenberger, K.H., Schuck, C., Shih, S., and Stephens, F. (1981). *Phys. Rev. Lett.* **46**:1383.

Nguyen, van Giai (1980). In *Highly Excited States in Nuclear Reactions*, eds. Ikegami and Muroaka, p. 682. Osaka: Osaka University.

Nguyen, van Giai and Sagawa, H. (1981). *Nucl. Phys.* **A371**:1.

Nguyen, van Giai, Bortignon, P.F., Zardi, F., and Broglia, R.A. (1987). *Phys. Lett.* **B199**:155.

Nguyen, van Giai, Chomaz, Ph., Bortignon, P.F., Zardi, F., and Broglia, R.A. (1988). *Nucl. Phys.* **A482**:437c.

Nilsson, S.G. (1955). *Mat. Fys. Medd. Dan. Vid. Selsk.* **29**(16).

Nilsson, S.G., and Ragnarsson, I. (1995). *Shapes and Shells in Nuclear Structure*, Cambridge: Cambridge University Press.

Nilsson, S.G., Tsang, C.F., Sobiczewski, J., et al. (1969). *Nucl. Phys.* **A131**:1.

Novotny, R., et al. (1991). *Trans. Nucl. Sci.* **38**:379.

Ormand, W.E. (1997). *Progr. Theor. Phys.* **124**(Suppl.):37.

Ormand, W.E., Bortignon, P.F., Broglia, R.A., Døssing, T., and Lauritzen, B. (1990). *Phys. Rev. Lett.* **64**:2254.

Ormand, W.E., Bortignon, P.F., Broglia, R.A., Døssing, T., and Lauritzen, B. (1990a). *Nucl. Phys.* **519**:61c.

Ormand, W.E., Bortignon, P.F., and Broglia, R.A. (1996). *Phys. Rev. Lett.* **77**:607.

Ormand, W.E., Camera, F., Bracco, A., Maj, A., Bortignon, P.F., Millon, B., and Broglia, R.A. (1992). *Phys. Rev. Lett.* **69**:2905.

Ormand, W.E., Bortignon, P.F., Broglia, R.A., and Bracco, A. (1997a). *Nucl. Phys.* **A614**:217.

Ormand, W.E., Bortignon, P.F., and Broglia, R.A. (1997b). *Nucl. Phys.* **A618**:20.

Pacheco, J. (1989). Ph.D. Thesis, University of Copenhagen (unpublished).

Pacheco, J., and Broglia, R.A. (1988). *Phys. Rev. Lett.* **61**:294.

Paul, P., and Thoennessen, M. (1994). *Ann. Rev. Nucl. Part. Sci.* **44**:65.

Pearson, J.M. (1991). *Phys. Lett.* **B271**:12.

Ponomarev, V.Yu., Vigezzi, E., Bortignon, P.F., Broglia, R.A., Colò, G., Lazzari, G., Voronov, V.V., and Baur, G. (1994). *Phys. Rev. Lett.* **72**:1168.

Puddu, G. (1993). *Phys. Rev.* **C47**:1067.

Puddu, G., Bortignon, P.F., and Broglia, R.A. (1991). *Ann. Phys. (N.Y.)* **206**:409.

Pühlhofer, F. (1977). *Nucl. Phys.* **A280**:267.

Quentin, P., and Flocard, H. (1978). *Ann. Rev. Nucl. Sci.* **28**:523.

Ramakrishnan, E., et al. (1996). *Phys. Rev. Lett.* **76**:2025.

Rapaport, J., et al. (1979). *Nucl. Phys.* **A330**:15.

Ritman, J., et al. (1993). *Phys. Rev. Lett.* **70**:533, 2659.

Sagawa, H., and Bertsch, G.F. (1984). *Phys. Lett.* **B146**:138.

Satchler, G.R. (1977). International School of Physics "E. Fermi," *Elementary modes of Excitation in Nuclei*, eds. A. Bohr and R.A. Broglia, p. 271. Amsterdam: North-Holland.

Schrieffer, J.R. (1964). *Theory of Superconductivity*. New York: Benjamin.

Shimizu, Y.R., Garrett, J., Broglia, R.A., Gallardo, M., and Vigezzi, E. (1989). *Rev. Mod. Phys.* **61**:131.

Shimizu, Y.R., Vigezzi, E., and Broglia, R.A. (1990). *Phys. Rev.* **C41**:1861.

Shlomo, S., and Youngblood, D.H. (1993). *Phys. Rev.* **C47**:529.

Slichter, C.P. (1963). *Principle of Magnetic Resonance*. New York: Harper and Row.

Snover, K. (1986). *Ann. Rev. Nucl. Part. Sci.* **36**:545.

Snover, K. (1993). *Nucl. Phys.* **A553**:153c.

Sommermann, H. M., Ratclif, K.F., and Kuo, T.T.S. (1983). *Nucl. Phys.* **A406**:109.

Speth, J., Cha, D., Klemt, V., and Wambach, J. (1985). *Phys. Rev.* **C31**:2310.

Stephens, F.S. (1985). *Frontiers in Nuclear Dynamics*, eds. R.A. Broglia and C.H. Dasso, p. 73. New York: Plenum Press.

Stephens, F.S., and Simon, R.S. (1972). *Nucl. Phys.* **A183**:257.

Stratonovich, R.L. (1957). *Dokl. Akad. Nauk. SSSR* **115**:1097.

Strutinsky, V.M. (1967). *Nucl. Phys.* **A95**:420.

Teruya, N., and Dias, H. (1994). *Phys. Rev.* **C50**:2668.

Thoennessen, M., Chakrabarty, D.R., Herman, M.G., Butsch, R., and Paul, P. (1987). *Phys. Rev. Lett.* **59**:2860.

Thouless, D.J. (1962). *The Quantum Mechanics of Many Body Systems*. New York: Academic Press.

Tveter, T.S., et al. (1996). *Phys. Rev. Lett.* **76**:1035.

Van der Woude, A. (1987). *Progr. Part. Part. Nucl. Phys.* **18**:217.

Veyssiere, A., et al. (1973). *Nucl. Phys.* **A199**:45.

Wang, S.J., and Cassing, W. (1985). *Ann. Phys.* **159**:328.

Weinberg, S. (1995). *The Quantum Theory of Fields*, Vol. 1. Cambridge: Cambridge University Press.

Weizsäcker, von, C.F. (1935). *Z. Phys.* **96**:431.

Wilkinson, D.H. (1956). *Phil. Mag.* **1**:379.

Yoshida, S. (1983). *Suppl. Progr. Theor. Phys.* **74–75**:142.

Yoshida, S., and Adachi, S. (1986). *Z. Phys.* **A325**:441.

Yoshida, S., and Adachi, S. (1986a). *Nucl. Phys.* **A457**:84.

Youngblood, D.E., et al. (1977). *Phys. Rev. Lett.* **39**:1188.

Index

H

Printed in the United States
by Baker & Taylor Publisher Services

Printed in the United States
by Baker & Taylor Publisher Services